Foreword

The first edition of this book was published in 1963, when traditional procedures in civil engineering were well established and widely accepted. When the 1979 edition was published the Drafting Committee could still write about 'the established organization and methods of working in the civil engineering industry'. These traditional procedures are based on the use of conditions of contract published by the Institution of Civil Engineers, in which the Engineer has a contract with the Promoter of a project to design and supervise, and the Contractor has a separate contract with the Promoter to construct the project works.

Since 1979 the traditional procedure has been challenged and there has been a search for other procedures for getting projects built, with the aim of producing better value for money in completing projects on time and within budget costs. The philosophy in this book is based on the traditional procedure, but reference has been made to the main alternative procedures as the civil engineer should be fully acquainted with the variety of contract strategies now in use in construction projects. It is emphasized that deliberate and careful consideration should be given to the most appropriate contract strategy for any particular project.

Speculation on the future of the industry is outside the scope of this book, but the procedures currently being used and those recently introduced, such as management contracting, are described. Thus it is hoped that the reader will at least be made aware of present trends and some of the difficulties which these are designed to overcome. If the contents appear to be conservative, it is because the members of the Drafting Committee believe that untried systems should be introduced only with caution. It is likely to take a number of years for a new system to be thoroughly tried and tested. Whatever system is used, it is most important that the responsibilities of the various parties are defined clearly.

The advice given is regarded as good practice and is for general guidance only. It may conflict with the provisions of a particular contract, in which case the latter, being legally binding, must prevail.

Generally Promoters are becoming more involved in the various stages of a project, particularly in the construction stage, and the independence of the Engineer on which the traditional system is founded is increasingly threatened. Promoters and their auditors are more ready to question Engineers' decisions.

Competition between contractors on a greatly reduced national work-load in recent years has resulted in a rigorous testing of not only the contract conditions but also the traditional procedure. Competition has been extended to the professions and consulting engineers have not been exempt. The long-term effects of such competition on professional performance and on the end product of the Promoter have yet to be seen.

Since publication of the 1979 edition of this book standard forms of contract have proliferated and changing technology has required more complex specifications, often making greater demands on quality of materials and workmanship. Failures inevitably receive much publicity and disputes, arbitration and litigation appear to be on the increase.

A flow chart has been included in this edition to give a simple picture of relationships involved in the building of a civil engineering project from inception to commissioning and hand-over to the Promoter for operational use and maintenance.

The importance of safety in construction is emphasized, with particular reference to the health and safety at work legislation. It is the responsibility of all engaged in construction, from project inception to final hand-over to the Promoter, that safe systems of working are set up and adhered to with quality standards assured for the completed project.

Although this book is applicable primarily to work in the UK, mention is made at the end of most chapters of special factors applicable to overseas work.

Chartered civil engineers play a vital part in every step of the system. As professional men serving the material needs of mankind it is their duty diligently to apply their training, experience and integrity of purpose to the operation of the contract strategy adopted by the Promoter for each particular project within the systems outlined in this book.

An increasing number of women are entering the industry and they will doubtless play a growing part in the future. The use of the male gender in this book is entirely for convenience and should be taken to imply both men and women.

Civil engineering procedure

Thomas Telford, London

1986

Published by Thomas Telford Limited, P.O. Box 101, 26–34 Old Street, London EC1P 1JH

First edition 1963
Second edition 1971
Reprinted with amendments 1976
Third edition 1979
Fourth edition 1986

Drafting Committee (1963 edition)

H. D. Morgan, MSc(Eng), FICE (Chairman)
Ralph Freeman, CBE, MA, FICE
R. le G. Hetherington, OBE, MA, FICE
W. K. Laing, MA, FICE
Colonel S. M. Lovell, OBE, ERD, TD, FICE
William MacGregor, DSc, FICE
J. H. W. Turner, BSc, FICE

Drafting Committee (1971 edition)

Sir Kirby Laing, JP, MA, FICE (Chairman)
D. C. Coode, FICE
Colonel S. M. Lovell, CBE, ERD, TD, FICE

Drafting Committee (1979 edition)

D. G. M. Roberts, MA, FICE, FIMechE (Chairman)
J. C. McKenzie, MAI, FICE
M. C. Purbrick, BSc(Eng), FICE

Drafting Committee (1986 edition)

J. V. Tagg, FICE, FIStructE, FRICS, FCIArb (Chairman)
F. Grover, BSc(Eng), ACGI, FICE, FIHT
P. H. D. Hancock, MA, FICE, FBIM
T. W. Weddell, BSc, DIC, FICE, FIStructE, ACIArb
C. J. Outhwaite, BSc, MICE

British Library Cataloguing in Publication Data
Civil engineering procedure.—4th ed.
1. Civil engineering 2. Engineering—Management
I. Tagg, J. V. II. Institution of Civil Engineers
624′.068 TA190

ISBN: 0 7277 0363 3

Typeset by M.H.L. Typesetting Limited, Coventry
Printed and bound in Great Britain by Billing and Sons Ltd, Worcester

Definitions

The 'Promoter' initiates the project and is responsible for providing funds for its execution. He is defined as the 'Employer' in the *ICE Conditions of Contract*, and in other forms of contract he is also referred to as the 'Purchaser'.

The 'Engineer' is appointed by the Promoter to have overall engineering responsibility for the investigation and design of the project, and to supervise its construction.

The 'Contractor' is the organization or individual entrusted with the construction of the Works.

The 'Contract' is the agreement entered into between the Promoter and the Contractor for the execution of the Works.

The 'Contract Documents' usually include the Form of Tender, the Conditions of Contract, the Specification, the Bill of Quantities, the Drawings, the Form of Agreement and the Performance Bond.

The 'tender' is the Contractor's offer to undertake the Works in accordance with the terms of the Contract Documents.

The 'Engineer's Representative' (usually called the Resident Engineer) represents the Engineer on the site of the Works.

The 'Agent' represents the Contractor on the site of the Works.

The 'Works' comprise all those items which are to be constructed under the terms of the Contract, i.e. all Permanent and Temporary Works.

'Temporary Works' (which are generally removed on completion of the project) are those items which are built to facilitate the construction of the Works.

Contents

Introduction

The procedure from initiation to completion of a project is conveniently illustrated by the flow chart in Fig. 1 which shows the three parties contractually involved in a project and the principal activities in their logical sequence.

In practice many of the activities overlap and therefore the sequence is seldom as simple as that shown. The Contractor is generally paid in stages, for work done at agreed payment times. Sometimes design and construction develop more or less in parallel, and it is not unknown for construction to start before a contract has been signed or even before the Promoter has acquired all his land and raised all his funds. Procedures and contractual methods vary with time and circumstances; reference is made in the section on contracts (p. 23) to the different types of contract in operation. The procedure shown in Fig. 1 is that of the traditional Promoter, Engineer and Contractor relationship where time allows the normal sequence of events to be followed.

There are several variants on this procedure. Three common ones are direct labour, as described in the section on means of execution (p. 22); all-in contracts, as described in the section on contracts (p. 27); and management contracts, as described in the section on contracts (p. 27).

Whatever may be the contractual relationship between the parties to a project, Fig. 1 embodies a principle of working that is well established in the UK and seems likely to remain valid for most purpose-built civil engineering projects. For example, the Engineer, whether a salaried employee of the Promoter or a consulting engineer, must retain his independence. The Contractor's engineer, who performs any of the functions shown in the Engineer column in Fig. 1 (all-in contract), has a similar problem. A prudently run direct labour organization will operate some kind of internal tendering procedure to keep a check on prices and provide a basis for cost control, and will appoint someone in the role of Engineer's Representative to ensure that his colleague who has the function of Agent executes the design correctly and to the required quality standards.

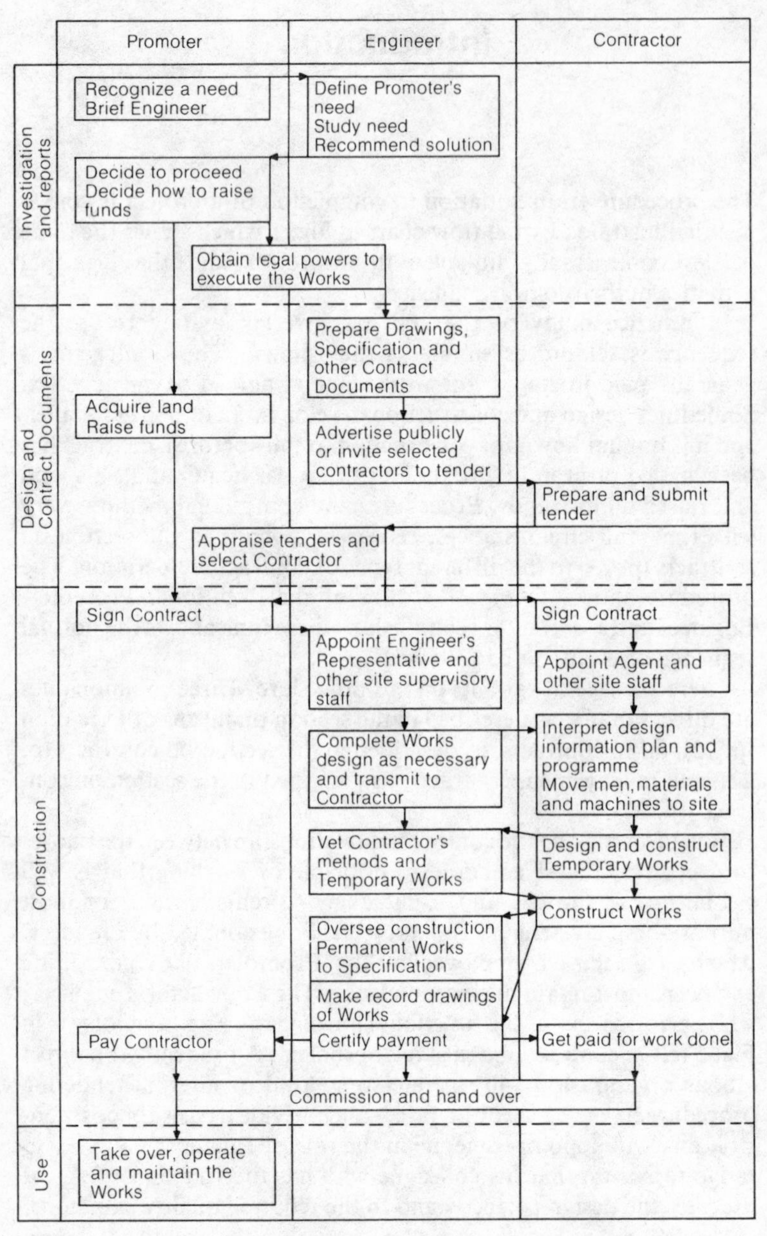

Fig. 1. Activities of the parties to a project

Promotion

Parties to a project

Traditionally there are three parties involved in a project: the Promoter, the Engineer and the Contractor.

The Promoter

The Engineer usually refers to the Promoter as the Client; the Contractor may refer to him as the Employer as he is called under some forms of contract. The Promoter may have virtually any legal status: he may be a sole trader, a partnership, a limited liability company or part of local or central government, or any other incorporated or unincorporated body.

The Engineer

The term 'the Engineer' refers to the person representing the organization which provides professional advice on the investigation for, and the design and construction of, civil engineering works. Consulting engineers usually practise in the form of partnerships but may practise as limited liability companies or as individuals. Alternatively the Engineer may be an employee of the Promoter. The position of the Engineer in relation to the auditors is outlined in a policy statement published jointly by the Institution of Civil Engineers and the Chartered Institute of Public Finance and Accountancy (see also Appendix 1).[1]

The Contractor

The firm which carries out the work of construction is the Contractor. Under the *ICE Conditions of Contract*[2] there is usually only one Contractor, sometimes referred to as the Main Contractor; on some works he may sublet or subcontract parts of the Works to specialist or other contractors who become known as subcontractors. Contractors and subcontractors are usually limited liability companies.

Authority for the project

Legal authority

Before a civil engineering project can be executed it is normally necessary to obtain various legal authorizations. These may, depending on the status of the Promoter, be by statutory regulation, legislation or parliamentary sanction. Public authorities and nationalized industries (e.g. British Coal and the Central Electricity Generating Board) usually have special powers delegated to them by central government.

Establishment of authority

The execution of most projects in the UK is dependent on the obtaining of planning permission and other statutory approvals. Applications for such have to define the scope of the work to be undertaken. Parliamentary sanction is required when the Promoter's constitution does not cover either the purposes for which the project is to be used or the construction of the project, e.g. the constitution of British Rail allows it to run railways but it was necessary to promote a private bill in parliament to obtain the power necessary to construct the east coast main line diversion around the Selby coalfield.

Projects which have to be approved under statutory regulations demand compliance with specific procedures as laid down in the regulations. These requirements call for special experience on the Engineer's part, and place a duty on the Promoter to initiate investigations in good time to allow all economic and other considerations to be given the attention they deserve. Lack of appreciation of the time required for proper investigation has been the cause of rejection of some projects, or of much additional expense which could have been avoided.

Promoters

Central government

Contracts are entered into with the Government through its departments. Under such contracts in the UK payment is made out of funds provided by vote of Parliament; expenditure in excess of this vote, although sanctioned by a department, is subject to a supplementary vote. The Contractor is not under any obligation to enquire whether or not a department is contracting beyond the funds voted by Parliament. The Government is bound in contract by any agreement made on its behalf by an agent having authority to enter into that contract. The role of each department and what type of work it is authorized to promote is set down by the Government of the day.

Local government

Local authorities acting under powers granted in charters and acts constituting them and under general or special acts governing their procedures and applications of funds can enter into contracts and raise funds for payments due under them. A contract validly entered into carries an implied undertaking that the authority possesses or will collect the requisite funds; lack of funds is no defence to an action for payment. For example, a local authority's liability will not be abrogated by the refusal of a government department to grant a loan sanction, because the local authority must be deemed to be ready to raise the funds required by, if all other means fail, levying supplementary rates. Local authorities are subject to the ordinary legal liabilities as to their powers to contract and their liability to be sued.

National or public industries or corporations

Statutory boards, corporations, basic industries and utilities (e.g. railways, postal services, coal mining, electric power, water supplies and some port authorities) are generally publicly owned. Their powers are laid down in the particular acts constituting them. The legal position of such bodies and of their officials is usually as stated for local authorities.

Incorporated companies

An incorporated company (i.e. a private or public limited liability company) can enter into contracts within the purposes of its memorandum of association or within the powers prescribed by any special act incorporating the company or any other act granting it powers for a specific purpose.

Other bodies

A contract entered into with a body of individuals not being a trading corporation (e.g. a partnership or a club committee) must be considered on its merits in respect of the authority of one or more of the individuals to bind the rest in personal liability for the purpose of the Contract and for making payments due under it. The safest course to adopt is to enter into contract with a sufficient number of members of the body to ensure that their collective financial status is adequate having regard to its value. Suitable references should be obtained, and inquiries made in any cases of doubt, especially before entering into a contract with a private individual.

Responsibilities and duties of the Promoter

The Promoter's responsibilities and duties are normally written into the memorandum of agreement with the Engineer and the contract for construction and are generally

(*a*) to define the functions that the project is to perform

(*b*) to provide information and data held by him and required by the Engineer
(*c*) to obtain the necessary legal authority to allow construction of the project
(*d*) to obtain money to finance the project
(*e*) to acquire the necessary land.

Project manager

The Promoter may decide to appoint a project manager. He may be a salaried employee or a consultant experienced in developing and controlling major projects from feasibility study through to commissioning. If the project is large enough or complex the project manager should appoint a team to cover the various engineering disciplines involved, or should arrange for their appointment by the Promoter.

Engineers

Selection of the Engineer

The Promoter may, depending on the type and extent of the project, decide to utilize the services of an engineer in his employment or may engage consulting engineers to provide engineering advice to enable him to assess the feasibility and/or relative merits of various alternative schemes to meet his requirements. Advice regarding suitable consulting engineers and their methods of engagement and working can be obtained from the Association of Consulting Engineers.

The selection of consulting engineers should start with an assessment of the qualifications and experience of possible contenders. Final selection may be by inviting proposals or competitive fee offers for comparison. Whenever the Promoter engages a consulting firm for a specific project a formal agreement should be completed between the two, which sets out the duties and responsibilities of each party and the fees and expenses to be paid. The Association of Consulting Engineers' *Conditions of engagement*[3] give guidance on these matters.

Status and qualifications

It is a matter of importance that the person appointed as the Engineer should have qualifications and experience appropriate to the project. Corporate membership of the Institution of Civil Engineers is recognized as an appropriate qualification for any position of responsibility in civil engineering. Corporate members of the Institution comprise Members who, having received appropriate education, training and experience, have been elected as such, and

General contractors are those who . . . are able to undertake responsibility for the construction of the whole of a project

Fellows who are senior engineers transferred to that class, having held positions of major responsibility on important engineering work for some years. The Engineer may either have official status in government or public service, public company or private enterprise, or be a senior member of a firm of consulting engineers and who, from his established reputation, is considered well qualified to advise on the development of a particular project. Corporate members of the Institution, being subject to the by-laws, regulations and rules of professional conduct of the Institution, are expected to be fit for positions of trust.

Duties

It is usual for the functions of the Engineer to include the development, design and technical direction of the Works, preparation of specifications, bills of quantities and other contract documents. Also, during the execution of the Works, his duties include the inspection of materials and workmanship and a variety of administrative duties such as the measuring and valuing of the work and the pricing of varied work.

Association of Consulting Engineers

The Association of Consulting Engineers was formed in 1913 to promote the advancement of the profession of consulting engineering (not exclusively civil engineering) and to ensure that its members advising on engineering matters are fully qualified engineers in their respective fields and at the same time willing to be bound by a strict code of professional rules. The Association also provides for the information of the public a list of independent engineers who are not directly or indirectly connected with any commercial or manufacturing interests which might affect their professional judgement. It affords a means by which the profession can confer with government and public departments as well as other professional bodies, and recommends conditions of engagement which, together with the professional rules, are generally accepted as the standard for the profession.

Contractors

Civil engineering contractors may be broadly classified under two headings: general contractors and specialist contractors.

General contractors

General contractors are those who, on account of their knowledge and experience, are able to undertake responsibility for the construction of the whole of a project.

Specialist contractors

Those who confine their activities to selected classes of work are referred to as specialist contractors. In some cases their operations are covered by patents, but generally they rely on special resources and experience. This specialization enables them to carry skilled staff and plant particularly suited to their work. The introduction of new processes and methods of construction is often due to the activities of such contractors and their employment can be of economic advantage to both Promoters and general contractors. A specialist contractor may perform his work by subcontract to a general contractor who will act as co-ordinator. Alternatively the specialist contractor may perform his work by direct contract with the Promoter.

Federation of Civil Engineering Contractors

The Federation of Civil Engineering Contractors was formed in 1919, its aims being to further the interests of its members, to establish amicable arrangements and relations between its members and their work people and to regulate wages and working conditions in the industry, to maintain a high standard of conduct, to combat unfair practices and encourage efficiency among its members, and to settle and secure the adoption of standard forms of contract embodying equitable conditions.

Overseas

Authority for the project

Engineers undertaking projects overseas should make themselves thoroughly familiar with local requirements and customs. These can vary considerably from country to country.

Promoters

In addition to the types of Promoter already mentioned, bodies such as the World Bank, the Asian Bank and various regional organizations promote civil engineering projects. Finance is also made available to poorer countries by the more wealthy, but this is usually done through promoting authorities in the country where the work is to be done. The authority and standing of Promoters should be checked in the country where the work is being undertaken.

Engineers

In some overseas countries an engineer may practise only if he is registered for that purpose in accordance with the laws of the country. Some governments specify that any foreign consulting engineer must form an association or partnership with a local firm or agent before he will be permitted to practise in their country. Engineers working overseas must be prepared to adapt to different customs, traditions and professional standards. Fee competition between consulting engineers is common, even in circumstances where it must result in a lowering of standards of work to the detriment of the Promoter.

International organizations

The Association of Consulting Engineers is a member of the Fédération Internationale des Ingénieurs-Conseils. This is an organization of consulting engineers' associations in a number of countries throughout the world which have similar interests to those of the Association of Consulting Engineers. Similarly, there is an international organization of contractors' federations – Fédération Internationale des Entrepreneurs Européens de Bâtiment et des Travaux Publics – with which the Federation of Civil Engineering Contractors is associated. Other such organizations exist in other parts of the world (e.g. the International Federation of Asian and Western Pacific Contractors' Associations, the Inter-American Federation of the Construction Industry and Associated General Contractors of America).

Investigation and reports

Need for investigation

The Promoter's need for advice

Having appointed or employed an Engineer, the Promoter may want him to assess the project requirements and to devise and present alternative schemes for his consideration, although occasionally he may know in general terms what he requires and will need to have only the details determined or developed. Sometimes he may want a particular phase or aspect investigated to provide the necessary information on which to base a decision on whether or not to proceed with a project, or to acquire finance or parliamentary or other authority for its implementation. In other cases he may wish to ascertain the effect a project will have on the environment (this type of study is known as environmental impact analysis). Although the engineer appointed for the investigation need not be the engineer who will design the Works nor become the engineer who will supervise the Works, the continuity achieved if this is so is usually advantageous to all parties.

Nature of the investigation

On appointment, the Engineer's first tasks are to become acquainted with and to review the Promoter's requirements, and then to agree a brief for the investigations. The Engineer as principal professional adviser should enjoy the confidence of the Promoter and be given timely knowledge of all the requirements so far as they can be defined. Normally the Engineer will inspect the site and consider all physical, financial and political constraints, devise possible solutions and compare them. Then he will submit a report to the Promoter summarizing his findings and conclusions and, usually, recommending a preferred solution with an estimate of its capital and likely maintenance costs.

Importance and extent of investigation

The Engineer's report on his investigations should recommend the basis for the economic design of a project.[4] It should therefore review the main design choices and the major problems which may affect construction. These should therefore have been investigated thoroughly by this stage. Investigations and reports cost only a small fraction of the total cost of construction, so it is false economy to restrict the scope or thoroughness of the Engineer's investigations and design studies.

The Engineer's terms of reference

Scope of work statement

The Engineer's duties in respect of investigation and reports should be defined in a scope of work statement as part of his brief. His brief is usually initiated by the Promoter, but the final version should be agreed between the Promoter and the Engineer so that full advantage can be taken of the latter's specialist knowledge and his experience of similar studies. In preparing the document the Promoter must state his long-term and short-term objectives and he must indicate all the constraints of which he is aware. The scope of work statement should also indicate the nature and details of the information to be presented in the Engineer's report and the supporting documents, what existing information can be made available, the time available for preparation of the report, and whether interim or progress reports will be required. Site conditions and their possible exploitation often present risks to the costs of projects which are not realized by Promoters. In such cases the scope of the basic objectives may be a matter for exploration and advice by the Engineer before the brief for the project is finalized.

Flexibility and consideration of alternatives

The scope of work statement should permit flexibility so that the Engineer can consider new developments that may occur while the investigation is in progress. The extent to which the Engineer may consider alternative solutions should be agreed. It is generally advisable for the scope of work statement not to indicate a preferred scheme, as the value of a report can be seriously diminished if it can be inferred that any matter has been prejudged. This is particularly true when environmental issues are involved.

Types of investigation or study

The extent, nature and detailed content of the Engineer's study will vary according to the value and complexity of the project, the number of solutions to be considered, and the nature and number of the decision-making processes involved before the scheme can be constructed. For some projects, a single study with the preferred scheme and its estimated cost outlined in the report will suffice; others will require a series of separate studies, each more detailed than the last. Similarly, some studies will be undertaken wholly by engineers, whereas others will be the work of a team of specialists of

This type of study is known as environmental impact analysis

many disciplines. The studies required to develop a preferred scheme for a major project may be some or all of the following

(*a*) a study which assesses the requirements of a particular sector of a country or industry in order to identify individual projects for investigation
(*b*) a pre-feasibility study which investigates whether or not there is likely to be a viable demand for the project to be studied, whether or not the required resources of implementation would be available within acceptable cost limits, and whether or not a feasibility study would be justified
(*c*) a feasibility (or pre-investment) study which involves preliminary surveys to investigate technical and economic viability, an estimate of capital and operating costs, and other information to enable the Promoter to decide whether or not he should try to finance the project; it does not include detailed design, but may include some outline design
(*d*) a master plan which is a long-term development programme, and which generally indicates how construction and expenditure can be phased
(*e*) an environmental impact study which considers the effect the proposed development would have on the environment
(*f*) a geotechnical study which investigates the nature of the ground where works may be constructed and assesses possible problems
(*g*) a project study which develops the preferred scheme in detail, especially its technical aspects; this is sometimes known as the final design study
(*h*) a financing study: in most cases the Promoter will make the necessary financing arrangements himself, but exceptionally the Engineer may be required to identify possible sources of finance and to assist the Promoter in establishing the arrangements for the provision and repayment of the funds needed, in which cases a special report on the financing arrangements will be needed.

Nature of investigations

Technical investigations

Technical investigations are required to resolve the engineering aspects of a project. Further investigations will be required on projects likely to affect the environment; economic, social or environmental studies may need to be undertaken in such cases.

Economic and social investigations

The need for economic, demographic or other studies depends on the type of project and its Promoter. Although they will generally be carried out under the direction of the Engineer, they will usually be performed by a specialist in another discipline. The team for any large study may include several non-engineering specialists.

Use of existing information and plans

For all investigations and surveys full use should be made of existing available data, and topographical maps and marine charts where these exist. It may be necessary to carry out additional studies to supplement the available information and, in the case of topographical investigations, to undertake land, air or hydrographic surveys to make possible the preparation of accurate plans of the sites considered for the construction of the Works.

Site exploration

Thorough site exploration is fundamental to the economic design, efficient planning and execution of any project. Its importance cannot be over-emphasized. Existing topographical and geological maps and data should be studied, and thorough investigations should be carried out under the direction of the Engineer. Special investigations (e.g. by resistivity and seismographic surveys), soils investigations and so on should be undertaken by specialists who should provide borehole logs, samples and laboratory test results. Their reports should include an interpretation of results by a geotechnical expert, but the Engineer must make the final assessment of the results in so far as they may affect the design and specification of the Works.

Physical data

Among the items that may require investigation are the conditions of access to the sites both for permanent use and also during construction, statutory or local authority restrictions on the sites or the special rights of adjacent landowners, and the availability of services such as drainage, sewerage, water, electricity and gas. The source and transportation of materials for construction and operation and the disposal of waste material should be studied, and data on the climate and the incidence of storms and flooding collected. The existing ambient noise levels may have to be measured and studies undertaken to predict, and restrict as necessary, the noise levels during and after construction.

Models

There are four principal types of model: experimental, mathematical, computergraphic and physical. Experimental and mathematical models are used in complex problems to provide

The team for any large study may include several non-engineering specialists

information for the assessment of the effects of various design options. Physical and computergraphic models can be valuable for planning the layout of projects, studying the use of space, and deciding on the methods and sequence of work in construction. Physical models are valuable for discussing proposals with a Promoter's staff and also for training those who will take over the operating activities of the completed project.

Public participation

The execution of most civil engineering projects affects the public in some way. Although the general public benefit from most projects, there are usually some people who suffer. For example, the land-owner whose land is compulsorily acquired for a new road scheme, and the householder who experiences noise nuisance as a result of the construction of a new airport runway, may both reasonably have objections to a proposed scheme. Attempts have been made to involve the public in the decision-making process at an early stage. Frequently the problem and possible solutions are publicized, and the public are encouraged to express their views and preferences, which are duly taken into account during the feasibility and planning stages. Involvement of the public in the planning stage of a project can reduce objections and eliminate the need for a public inquiry; it can also produce constructive ideas to the benefit of the project.

Outline programme

An important part of an investigation is the preparation of an outline programme for the detailed design and construction of the Works. In formulating this the Engineer must take account of the availability of finance, labour, materials and feasible methods of construction and Temporary Works: each of these factors may influence the others. The outline programme must be realistic but it should not impose unnecessary restrictions on tenderers or the Contractor.

Cost estimates

The Engineer is usually required to prepare estimates of the capital cost of proposed schemes, of the land, buildings, plant and machinery involved, and of the operating and maintenance costs of the project. Some of these estimates may have to be obtained from or confirmed by outside sources. Capital and annual cost estimates will have to be drawn up so that the economic viability of alternative schemes or sites can be assessed and the costs to the Promoter of implementing the schemes ascertained. When comparing alternatives economic techniques such as discounted cash flow and net present value should generally be adopted.[4] It is essential to state the month and year or other basis on which cost estimates have been prepared, and in financial plans to indicate the estimated actual expenditure in future years and the rates of inflation assumed.

The Engineer's report

Objective

On completion of his investigations the Engineer should present to the Promoter a report summarizing his investigations and conclusions. Normally he will outline the scheme which, in his opinion, is best suited to the requirements (which will usually, but not always, be the least costly of the feasible options), and give forecasts of costs and financial viability to help the Promoter visualize the project as a whole.

Presentation

The style and form of the Engineer's report should vary according to the Promoter's needs, the nature of the project and the purpose for which the project was commissioned. As a general rule the report will review the investigations undertaken, compare feasible options on technical, economic and financial grounds, and make appropriate recommendations on which the Promoter can base his future decisions. The report should be written in a simple and unambiguous form so that it can be readily understood by the Promoter and others who may refer to it, who may neither be able to appreciate the technical detail nor have the time to absorb it. The report must contain sufficient technical information to convince any other engineer or specialist of its facts and of the soundness of its judgements.

Form of report

The report should set out in a clear and concise manner what the Engineer has been asked to do, what he has done and his conclusions. It is often convenient to summarize the salient points in one chapter and to include the technical matter in appendices or in separate volumes. In most cases the main body of the report should include

(*a*) the terms of reference
(*b*) a statement of the problem, its importance, scope and history
(*c*) the basic assumptions, data or trends on which the Engineer's

investigations have been based, and his reasons for adopting these

(*d*) details of the investigations carried out

(*e*) the design and other criteria adopted and the reasons for their adoption

(*f*) a comparison of the options considered and the reasons for the rejection of non-preferred schemes

(*g*) cost estimates of the schemes

(*h*) a description of the recommended scheme

(*i*) proposals for the organization and management of the design and construction of the recommended scheme, advice on the contract arrangements and perhaps the raising of finance

(*j*) conclusions and recommendations.

Proceeding with the project

The next stage

If, having considered the Engineer's report, the Promoter decides to proceed with the project he must take steps to obtain the legal authority to construct the Works. In all cases it is necessary to obtain planning permission or its equivalent as part of the procedure, and to seek comment from adjacent land-owners or other interested parties who may have objections to the project. Having obtained this the Promoter can instruct the Engineer to proceed with the design of the project and the preparation of tender documents. This work will usually proceed concurrently with the procurement of finance and the purchase of the necessary land.

Public inquiries

Whereas the questions of need and national policy are generally matters for Parliament, proposals for a particular project are of concern mainly to those who live in the area. For many types of project provision is made by statute for holding a public inquiry, which is a public hearing of objections to a particular proposal. The public inquiry is chaired by an independent inspector (called a reporter in Scotland). At the hearing the Promoter puts his case for the scheme, calling witnesses in support as necessary. The objectors are then heard. Parties are frequently represented legally, and expert witnesses are called if the scheme is of a highly technical nature. As part of his duties, the inspector visits the site during the proceedings.

After the public inquiry, the inspector writes a report which he submits to the appropriate government minister. This report contains an outline of the evidence given at the inquiry, the inspector's findings of fact and his recommendations. In due course the government minister makes his decision, which does not necessarily have to be in accordance with the inspector's recommendations. Appeal against the minister's decision is possible on legal grounds but not on the merits of the decision.

Overseas

Investigations and reports for projects overseas are fundamentally similar to those for projects in the UK. However, knowledge of local materials, labour and skills, available plant, climatic conditions and social customs should be taken into account.

Design and preparation of Contract Documents

Preparation by the Engineer

Design

To enable the Engineer to prepare designs for the Works and to provide adequate information on which contractors can be invited to tender, further site investigations are usually required. Laboratory tests or model tests may also have to be undertaken, and the approval of outside authorities obtained for certain aspects of the proposed Works. The Engineer must give attention throughout to the aesthetic and environmental aspects of the design, and should be free to seek other advice as he wishes on these and other subjects. Traditionally in the UK the experience of contractors was brought in only at the tender stage, but consultation with contractors during the design stage is now advocated by some Promoters who wish to make sure that the design is suitable for safe and economic construction.

Specialist firms

The Engineer may need to employ specialist firms for consultation and design on certain parts of the Works. The design of specialist work may be delegated to specialists who may later, with the Promoter's approval, become nominated subcontractors for the detailed design and construction stages of the project. Equitable arrangements for payment for specialist design work must be made between the Engineer and the specialist consultant or contractor. The Engineer should retain ultimate responsibility for all specialist design work.

Outline designs

In some cases outline designs, giving more information than was contained in the Engineer's initial investigation report but less detailed than tender designs, may be required for approval by the Promoter. The need for such designs should be stated in the agreement between the Promoter and the Engineer.

Scope of tender designs

Ideally, detailed designs of all the Works should be completed before tenders for construction are invited. In some cases this is not practicable because of the urgency of the Works, and the drawings on which tenders are invited are supplemented by a series of further drawings issued by the Engineer during construction.

The process of design itself is outside the scope of this book, but the more complete the drawings, specifications and bills of quantities are at the time of the call for tenders, the better the tenderers will understand what is required, the more accurate will be their prices, the smoother the subsequent execution of the work will be and the lower the cost. Certain information can be known only when construction is under way, and in such cases redesign or supplementary design work during the construction phase may be unavoidable. Examples include excavations in ground which proves to be different from that inferred from the information available at the time of design, and structures housing machinery the precise details of which are unknown at the design stage.

Time

The Engineer should advise the Promoter of the time needed to prepare designs and Contract Documents, as shortage of time at this stage is highly likely to lead to delays and additional expenditure. An early decision should be taken as to the type of contract to be used for construction, as this will affect the form of the documents the Engineer has to prepare.

Means of execution

Types of contract

Civil engineering construction is usually executed under a contract entered into between the Promoter and a contractor. Contracts may be classified as

(*a*) admeasurement contracts, in which the basis of payment to the Contractor is a priced bill of quantities and/or priced schedule of rates
(*b*) lump sum contracts
(*c*) cost reimbursement contracts[5]
(*d*) all-in contracts
(*e*) management contracts.[6]

Direct labour

As an alternative to execution by contract the Promoter may decide to use the method known as direct labour. Under this system the Promoter uses labour already available within his own organization or recruited specifically for the project. He undertakes all the responsibilities of management and risks of construction, and pays all wages and expenses as they are incurred. There may be no contract in the normal sense, unless the Promoter requires the work to be undertaken in competition with contractors.[7]

Contracts

Forming a contract

Under English law a contract is formed when an offer (e.g. a tender) by one party is unconditionally accepted by the person to whom it is made – the second party. The word 'unconditionally' implies that full and complete agreement has been established between the parties. An acceptance accompanied by ifs and buts or provided that does not form a contract but constitutes a counter-offer, which in its turn requires unconditional acceptance by the first party. This acceptance must be communicated to the party making the offer – silence does not mean consent. If it is sent by post it is considered to have been communicated (and acceptance is considered complete) at the moment it is posted; it is not prejudiced by loss or delay in the post. Proof of posting is therefore important. To be effective, an acceptance must be made within any limit of time stated in the offer, or, if none is stated, within a reasonable time.

Elements essential to a valid contract

In addition to the agreement evidenced by offer and acceptance there are other essential elements of a contract. Some affect every contract, but others are not likely to be met in the normal run of engineering contracts such as are of concern in this book. For the sake of completeness all are mentioned. Some of the requirements cause a contract to be void automatically if they are not fulfilled. Others result in a contract which can be voidable (or partly voidable) as of right by one of the parties – usually the one who has been aggrieved or put at a disadvantage by the non-fulfilment. The main requirements are

(*a*) an intention by the parties to create a legal relationship between them
(*b*) a genuine consent of the parties
(*c*) legal capacity of the parties to enter into a contract
(*d*) legality of the objectives of the Contract
(*e*) valuable consideration passing between the parties for simple contracts (consideration is not necessary for a contract under seal).

Basis of payment

In admeasurement and lump sum contracts the Contractor is paid for the work done in accordance with rates and/or prices tendered by him beforehand, whether in competition or otherwise. In cost

. . . silence does not mean consent

reimbursement contracts, he is paid the actual ascertained costs incurred by him under predefined heads, plus a fee to cover his administration and profit. In some cases a bonus and a penalty based on the time for completion of the Works may also be introduced and similarly based on the actual cost of the Works, as in target contracts.

Bill of quantities contracts

The bill of quantities contract is based on a detailed bill of quantities (computed by the Engineer's organization from the Drawings) which includes brief descriptions of the work to be undertaken, against each item of which the contractors tendering enter a unit rate or price. The tender total is the aggregated amount of the various quantities priced at the quoted rates. During the performance of the work the actual quantity executed under each item is measured and valued at the quoted rate. In the *ICE Conditions of Contract*[2] provision is made for the adjustment of rates for varied or additional work and for the fixing of new rates by the Engineer.

Schedule of rates contracts

The usual form of the schedule of rates contract is a list of items of work covering the operations which a Promoter may want done. No quantities are given. Sometimes tenderers are invited to affix rates to the items; sometimes they are invited to quote a percentage to be added to or deducted from rates previously entered by the Engineer. The Contract usually relates to a defined location and to a stated period of time. The Contractor may be called on to undertake any of the items of work anywhere in that location during the term of the Contract.

Another form of this type of contract is that applied to one specific project where it is essential that construction is started without delay. In this case a short list of the most important items of work is prepared, complete with representative quantities. These items are priced by the Contractor and, if accepted, these prices become the basis for negotiating prices for subsequent items of work as they become necessary.

For all civil engineering work mutual trust and confidence between the parties involved is necessary to make the Contract work effectively; this is so particularly for schedule of rates contracts.

Lump sum contracts

A lump sum contract is one in which the Contractor undertakes to carry out all the work specified and shown on the Drawings for a tendered sum of money. The Contractor is responsible for the assessment of all the costs he will incur in fulfilling the specified requirements. The Contract will not usually provide for any adjustment of the quoted lump sum unless the requirements specified at the time of tender are altered. Payment is not made as a single sum other than for very small jobs, but is made in interim stage payments related to the amount of work done or at stated intervals during the progress of the Works.

Tenderers for such contracts may be asked to quote rates for items of work listed in a schedule for use by the Engineer for the fixing of prices for new or varied work ordered during the execution of the project.

Cost reimbursement contracts

For a contract in any of the forms set out below a comprehensive list is prepared of all the heads of expenditure, such as salaries, wages, insurance, materials, plant, tools and consumable stores. The Contractor is usually obliged to obtain competitive quotations for resources to be used before placing orders. The total expenditure incurred under these headings is the allowable (or prime) cost. A fee is agreed beforehand for the work to be undertaken by the general resources of the Contractor's organization, which will include head office, other overheads, and for profit. The sum of the allowable cost and this fee is the total amount paid by the Promoter to the Contractor.

(*a*) *Cost plus percentage fee contract*. The fee is a fixed percentage of the allowable cost.
(*b*) *Cost plus fixed fee contract*. The fee is a fixed lump sum, usually based on an agreed estimated cost range.
(*c*) *Cost plus fluctuating fee contract*. The fee is tied to the allowable cost by a sliding scale.
(*d*) *Target contract*. This form of contract is designed to encourage the Contractor to effect economy in cost. Usually a basic fee is quoted as a percentage of an agreed target estimate obtained from a priced bill of quantities. Provision is made for adjusting the target estimate for variations in quantities and design and for fluctuations in costs of labour and materials. The actual fee payable to the Contractor is obtained by an increase or reduction in the basic fee, as the case may be, of agreed percentages of the amount of saving or excess between the actual cost and the adjusted target estimate.

Contract documentation

In measurement, lump sum and cost reimbursement contracts the Contract Documents are prepared by the Engineer acting on behalf of the Promoter. They are used first to obtain tenders from contractors, and subsequently in the administration of the Contract.

All-in contracts

In an all-in contract the Promoter, through his own organization or through an Engineer he has appointed, states his requirements in broad and general terms only and invites contractors to submit complete proposals and terms of payment for the design and construction and, if required, the commissioning, operation and maintenance for a limited period, of the project. This procedure has been in use for many years in certain fields of engineering construction, as for instance in the chemical and oil industries, and for the design and construction of nuclear power stations.

To undertake such contracts some manufacturing and contracting firms have formed themselves into consortia, while often also retaining the services of consulting engineers, in order to command all the resources needed to design, construct and equip such projects completely and to offer a contract for so doing. Such contracts are often commercially favoured by Promoters, particularly overseas, but offers from contractors should be carefully vetted by an independent consulting engineer on behalf of the Promoter to ensure that the Promoter understands fully all aspects of the various offers.

Management contracts

Initially a management contractor will provide pre-construction services to the Promoter and his professional teams. The extent of these services will vary with the character of the project and the stage in a project's development at which a management contractor is appointed. A management contractor does not normally undertake any of the construction work on site. He provides management services to control and co-ordinate all site activities, which are let to other contractors on a competitive basis. Organizations which carry out construction work within a project administered by a management contractor are sometimes known as construction contractors, to avoid confusion with the role played by subcontractors in a conventional contract.

Several forms of management contracting have developed. They are distinguished by the range of activities that they are intended to cover and include

(*a*) management contracts
(*b*) construction management contracts
(*c*) design and management contracts

(*d*) project and management services contracts (as used in the process and offshore industries).

Since management contracting systems were introduced in the UK in the late 1960s they have been widely used in the building, process and offshore industries, but their use is still rare in civil engineering.[6]

Subcontracts

There are two types of subcontract: nominated and non-nominated. In the former, which are used generally for specialist work and plant installations, the Promoter or Engineer selects the subcontractor following subcontract tenders to the Promoter and instructs the Contractor to enter into a subcontract in terms determined by the Promoter or Engineer, but subject to the proviso that the Contractor is not bound to do so unless the subcontractor is prepared to accept responsibilities and obligations comparable with those imposed on the Contractor under the terms of the Contract. In non-nominated subcontracts, known as domestic subcontracts, the contract terms are agreed between the Contractor and the subcontractor; such subcontracts may be entered into only with the prior consent of the Engineer.

Serial contracts

Sometimes the Promoter may invite tenders for the supply of materials or the carrying out of work at intervals as required over a period of, say, 12 months. These tenders, called serial tenders, are standing offers which may be accepted by the Promoter as required. They are usually in the form of a schedule of rates, and are subject to minimum quantities of materials or work being required.

Mechanical and electrical plant contracts

Many projects involve the supply and installation of mechanical and electrical plant, the contractual arrangements for which must be decided by the Promoter and the Engineer, and co-ordinated with the Contract for the civil engineering construction. If both the civil and mechanical/electrical work proceed to programme, any of the arrangements mentioned below will be satisfactory, but if either party departs from the programme or meets difficulties which have repercussions on the other, serious contractual problems may arise. The contractual arrangements adopted for the supply and installation of the mechanical/electrical equipment must be considered carefully and regard paid to the circumstances. Sometimes an arrangement which is advantageous at the tender stage may be less so at the construction/installation stage, and vice versa. The usual arrangements are

(*a*) direct contract, where the Promoter enters into a separate contract with the mechanical/electrical contractor
(*b*) nominated subcontract, where the Promoter instructs the civil engineering contractor to enter into a subcontract with a mechanical/electrical contractor of the Promoter's choice
(*c*) non-nominated subcontract, where the civil engineering contractor selects and appoints the mechanical/electrical contractor (sometimes inviting quotations from an approved list, using a specification and subcontract document prepared by the Engineer).

Mechanical and electrical contractors normally work to one of the standard forms of contract prepared by the Institutions of Mechanical and Electrical Engineers,[8] which contain provisions suited to mechanical and electrical work but which are different from those in the *ICE Conditions of Contract*.[2] If the Promoter enters into a direct contract for the electrical/mechanical work there should be no problems in working to such conditions. If such work is subcontracted, the Contractor should allow in his tender rates and prices for any differences between the two sets of conditions.

Factors affecting choice of type of contract

Requirements of the Promoter

In weighing the merits of the different types of contract the wise Promoter will wish to know as accurately as possible before the Contract is placed the total expenditure he will incur and the time required for completion of the work. He will, in most cases, also wish to ensure that the work is done at the minimum cost compatible with satisfactory materials, workmanship and time. The extent to which these objectives are attained will depend largely on the quality and scope of the information embodied in the documents on which contractors are invited to tender. The better the information, the more precise will be the tendering and the more accurate will be the forecast of the final project cost.

Exceptional urgency

It is usual to invite competitive tenders for a contract in the expectation that this will make contractors plan to do the work efficiently and therefore at minimum cost. The value of time to the Promoter may sometimes be so great as to dictate a start of construction work before the detailed design is completed, or even concurrently with its beginning. If it is not possible to obtain competitive tenders, a cost reimbursement contract may have to be used. Such an arrangement gives the Engineer more scope in exercising discretion to amend design as construction proceeds, and this is often valuable in fields where technology is changing rapidly or where there are long lead times for the delivery of essential items of plant and equipment.

Contingencies and risks

Contingency sums are sometimes specified to be included in tenders for likely but unknown detailed items of work, to be expended at the direction of the Engineer. There are also unforeseen risks against which insurance cover must be provided. However much care and time is devoted to preliminary investigations, the contingent and unforeseen risks are usually potentially costly in civil engineering work. The uncertainty of occurrence or magnitude of risks invites insurance, and protection is usually best secured by the selection of a contractor who, by reason of his experience and reputation, can obtain cover against the risks at competitive rates in the insurance market.

The extent and nature of the liabilities imposed under an admeasurement contract will have a direct bearing on the contract price. If a particular risk is disproportionately large it is more economical for the Promoter to carry it. If the Promoter has reason to secure protection through the Contract against every conceivable construction risk this will inevitably result in higher tender prices. The Engineer is the best fitted of the Promoter's advisers to assess the effects of contractual liabilities and to judge how far their imposition or relaxation will be of economic value to the Promoter; in fact, a sound sense of engineering judgement based on experience may be far more telling than mathematically correct economics.

Financing a project

The method of financing a project may affect the choice of type of contract entered into; the Promoter should keep the Engineer fully informed of his intentions in this respect, so that the Engineer can weigh this factor along with all the others to be considered. In some cases, particularly for overseas projects, the Contractor is required to raise at least part of the finance for the project, for instance through a banking house. The Export Credit Guarantee Department of the Government provides insurance to cover short-term financial needs of Contractors and guarantee the payment of bad debts.

Inflation

Civil engineering contracts are generally let on a fixed price basis for short contract periods, the limits of which vary from time to

The method of financing a project may affect the choice of type of contract

time depending on the prevailing rate of inflation in the UK. Under fixed price contracts no adjustment is made to the Contractor's rates inserted in his priced bill of quantities or lump sum tender for subsequent variations in the costs of materials, labour or plant from those ruling at the time of tender. Contracts for periods above the limit set by the Government and the industry at the time include a clause in which a formula is adopted which will provide for increased payments to the Contractor to reflect the consequence of changes in basic costs. By this means most of the risk of inflation is borne by the Promoter. The Baxter indices for civil engineering and the Osborne indices for building are currently used in the UK.

Choice of type of contract

The choice between the types of contract thus resolves itself into a complex process of balancing technical, economic and contractual factors. In this, the wise Promoter will rely largely on the advice of the Engineer.

Tender documents

Information given to tenderers

The tender documents should give tenderers all available information which is likely to influence their tenders. These documents should therefore convey the clearest possible picture of the character and the quantity of the work, the site and subsoil conditions, the responsibilities of the Contractor, the terms of payment and any other conditions, technical or commercial, which may affect the execution of the work. Such information is vital to the tenderer to enable him to work out a viable plan of operation, make provision for the necessary Temporary Works, estimate the cost of the work, assess the finance required and evaluate the risks involved. A lack of adequate and reliable information at the time of tendering will lead to uneconomic tender prices and claims for additional payments during the execution of a Contract.

List of documents

The documents in connection with an admeasurement contract generally consist of

(*a*) instructions to tenderers, which do not usually form part of the Contract Documents
(*b*) the Form of Tender (with which may be associated the requirements for a tender bond, details of experience and financial resources)
(*c*) the Conditions of Contract

(*d*) the Specification
(*e*) the Drawings
(*f*) the Bill of Quantities
(*g*) data affecting the execution of the Works, including such things as a site investigation report, details of access, limitations of working hours, local conditions and other work being carried out on site, machinery, services and supplies which will be provided by the Employer for the Contractor's use; hydrographic surveys and special safety measures
(*h*) the Form of Agreement
(*i*) the Performance Bond (if required).

In lump sum, cost reimbursement and all-in contracts, some of these documents may not be required, depending on the circumstances. In the case of direct labour, although there may be no contract, documents (*d*) and (*e*) and maybe (*f*) will nevertheless be required.

Instructions to tenderers

The purpose of the instructions to tenderers is to direct and assist the tenderers in the preparation of their tenders, to ensure that they are presented in the form required by the Promoter and the Engineer. These instructions will vary on every project. Some of the more usual and important items frequently included are

(*a*) documents to be submitted with tenders, including a programme and a general description of the arrangements and methods of construction which the Contractor proposes for carrying out the Works, together with particulars of the legal and financial status and technical experience of the tenderer
(*b*) the place, date and time for the delivery of tenders
(*c*) instructions on visiting the site
(*d*) instructions on whether or not tenders on alternative designs will be considered, and if so the conditions under which they may be submitted
(*e*) notes drawing attention to any special Conditions of Contract, materials and methods of construction to be used and unusual site conditions
(*f*) instructions on completion of the Bill of Quantities and/or schedule of rates, and production of tender and/or Performance Bonds.

In the case of selective tendering the particulars of tenderers' legal and financial status and technical experience are required at a prequalification stage.

Form of Tender

The Form of Tender is the tenderer's written offer to execute the work in accordance with the other Contract Documents, and for such sum as may be ascertained in accordance with the Conditions of Contract and within the time for completion. Normally the tenderer must submit a tender which complies fully with the Specification, but in addition he may also submit an alternative offer.

Conditions of Contract

The Conditions of Contract define the terms under which the work is to be carried out, the relationship between the Promoter and the Contractor, the powers of the Engineer and the terms of payment. The imposition of Conditions of Contract which are biased in favour of either party can be uneconomical. It is recommended that the *ICE Conditions of Contract*,[2] agreed between the Institution of Civil Engineers, the Association of Consulting Engineers and the Federation of Civil Engineering Contractors, are used wherever practicable.

It is inadvisable to make modifications to the *ICE Conditions of Contract*, as the legal result may often be different from what is intended. Where alterations to meet the needs of a particular project are unavoidable, the addition of special conditions is preferable to the modification of standard clauses; attention should always be drawn specifically to such amendments and additions. The standard conditions should not be altered or added to without appropriate legal or expert advice.

There are other standard forms of contract used for civil engineering work, particularly for specialist work, such as the *ICE Conditions of Contract for Ground Investigation*.[9] Other frequently used main forms of contract are the Joint Contracts Tribunal standard forms of building contracts and the GC/Works/1 and 2 and Form C1001 used by all central government departments for their capital works contracts, with the exception of the Department of Transport which has mainly used the *ICE Conditions of Contract*. These and other standard forms are listed in Appendix 2. Equivalent terms used in some standard forms of contract are given in Appendix 3.

Specification

The Specification describes in detail the work to be executed, the character and quality of the materials and workmanship, and any special responsibilities of the Contractor that are not covered by the Conditions of Contract. It may also lay down the order in which various portions of the work are to be executed, the methods to be adopted, and particulars of any facilities to be afforded to other contractors.

Care should be exercised when drafting the Specification to avoid conflict with any of the provisions of the Conditions of Contract. Some organizations, such as government departments, nationalized industries and large private sector organizations, have their own standard specifications. If these are available for purchase and this is stated in the Contract Documents, they need only be referred to in the Contract Documents, and statements given of modifications for the particular contract.

Drawings

Ideally the Drawings should give details of all the Works. For many reasons this is not always practicable, but tenderers must be given sufficient information to enable them to understand what is required so that they can submit properly considered tenders. All available information on the topography of the site and the nature of the ground should be made available to tenderers, preferably by being shown on the Drawings.

Bill of Quantities

The Bill of Quantities is a list of items giving the quantities and brief descriptions of work. In conjunction with the other Contract Documents, it forms the basis on which tenders are obtained. When priced it affords a means of comparing tenders. When the Contract has been entered into, the rates in the priced Bill of Quantities are applied to assess the value of the actual quantities of work carried out. The Bill of Quantities should be prepared and measurements made in accordance with the procedure set out in the *Civil engineering standard method of measurement*,[10] unless other methods of measurement are preferred by the Promoter.

Form of Agreement

The Form of Agreement, if and when completed, is a legal undertaking entered into between the Promoter and the Contractor for the execution of the work in accordance with the other Contract Documents. It is not essential to have a Form of Agreement if there is a written acceptance by the Promoter of the tender submitted by the Contractor.

Performance Bond

The Performance Bond is a document whereby a bank, insurance company or other acceptable guarantor undertakes to pay a

This system . . . has serious disadvantages and is generally not recommended

specified sum if the Contractor fails to discharge his obligations satisfactorily. It is not always required by the Promoter.

Procedure for arranging a contract

Inviting tenders

The Promoter who decides to have work executed by contract may select the Contractor by open tendering or selective tendering. Occasionally he may prefer to negotiate a contract with a particular contractor. This procedure does not have the benefits of competition, but it may be appropriate in cases of urgency, or where the selected contractor's expertise is unique and the firm is well known to the Promoter. Useful information on tenders is given in reference 11.

Occasionally tendering takes place in two stages. During the first stage selected tenderers submit bids based on an outline design or, in the case of an all-in contract, on design proposals. The second stage comprises bids on a preferred design, or the development of a detailed design and negotiation of a contract with a tenderer.

Open tendering

Under open tendering, tenders are invited by public advertisement. This system may appear to provide maximum competition, but it has serious disadvantages and is generally not recommended. The Engineer may have the problem of deciding whether or not to recommend a tender which is not the lowest, and this may conflict with the standing orders of the Promoter. A large number of tenders may be submitted, including some from firms of inadequate experience or doubtful financial standing. The system also involves too many contractors in abortive tendering and hence wasted resources.

Selective tendering

The practice of inviting tenders from selected contractors eliminates the undesirable factors connected with open tendering. Tenderers may be selected from a standing list or on an *ad hoc* basis for each project. Promoters having a continuing programme of works frequently keep standing lists of approved contractors for various types and values of work. These lists are normally reviewed periodically to allow for changing circumstances. An *ad hoc* list may be compiled after public advertisement or from knowledge of suitable firms. The preparation of either standing or *ad hoc* lists is usually by means of a pre-qualification procedure, in which contractors invited to pre-qualify are asked to submit details of their relevant experience and other information. They are then assessed on such factors as their financial standing, their technical and organizational ability, their performance and their health and safety records. Pre-qualification may not be necessary for firms already well known to the Promoter or the Engineer.

The aim is to identify only those firms that are capable of carrying out the Contract satisfactorily. Normally the number of contractors invited to tender is not fewer than four nor more than eight. Major Promoters in the UK and international bodies such as the World Bank, the International Bank for Reconstruction and Development and the EEC have standard procedures for selecting tenderers.

EEC directives

Since 1973, procedures for awarding UK contracts valued at more than a million European currency units (approximately £700 000 at the third quarter of 1986) have been specified by EEC directives and are obligatory for government departments and local authorities. Their purpose is to remove discrimination against contractors on grounds of nationality alone. The main requirement is for details of contracts to be published in the *Official Journal of the European Communities*. Under the restricted tendering procedure, firms must be selected only from those who respond to the advertisement, and no advertisement may be published in the UK press before the date of posting of the details to the *Official Journal*.

Joint bids

For major or complex contracts, especially where the financial and other commitments are large, a group of two or more contractors may submit a joint bid. If successful, the group may form a joint venture or consortium in which each contractor retains his separate identity but is responsible jointly and severally with the other(s) for the due performance of the Contract. Alternatively the group may form a special company for carrying out the particular Contract. The Promoter will normally require to know the legal status of the group. Whatever the arrangement, all the members of the group should bind themselves by an agreement defining the responsibilities of each member and the resources in respect of finance, labour and plant which each is expected to provide.

Tender period

In preparing his tender, a contractor usually has to make extensive inquiries and he must be given sufficient time to do so, to reduce the risk of his tender being based on inadequate knowledge and leading to difficulties if he is awarded the Contract. A period of at least four weeks should be allowed for tendering; for large projects at least eight weeks is recommended. Enquiries made by tenderers to either the Promoter or the Engineer should be dealt with on a strictly formal basis, to avoid advantage being given to one particular tenderer. Explanations of and changes to the Contract Documents should be in writing and distributed to all tenderers.

Estimating

The system of estimating varies with different contractors. The essence of any system is that it should be an orderly, convenient and comprehensive means of ensuring that all expenditure is provided for without omissions or duplication. Normally different parts of the Works are priced by assessing the labour, plant and materials content and pricing these using appropriate historical cost data. Work intended to be subcontracted is priced by obtaining quotations from likely subcontractors. It is important in cost reimbursement contracts that cost is clearly defined. Guidance on these matters is given in reference 5.

Tender rates and prices

The rates and prices entered in the Bill of Quantities are those for which tenderers offer to undertake the various items listed in the bill. They are based on their estimate of the costs of each operation, together with amounts – sometimes proportioned over all the items – to cover overheads, site offices, head office expenses, financing, insurance, general contractual liabilities, profit, contingencies and so on. Alternatively some of these items may be in a section of the bill on preliminaries. If the Bill of Quantities is based on the *Civil engineering standard method of measurement*[10] tenderers may enter both fixed and time-related items in the method-related charges section of the bill, and include in the rates entered against the items of permanent work only those elements which are quantity-related.

Tender total

The multiplication (usually referred to as the 'extension') of each tender rate by the corresponding estimated quantity of work stated in the Bill of Quantities will give an amount for each item of work. The tender total is the figure obtained by the summation of all such amounts, including sums entered for preliminaries, method-related charges and other parts of the Bill of Quantities. It is normally the main figure used in comparing tenders.

Difference between tenders

The totals of tenders from equally competent contractors may differ considerably, not so much because routine operations are differently priced, but because of different tendering strategies, construction methods and techniques, and because of differing values attached to risks and to methods of dealing with them. Profit margins may vary according to firms' tendering policies at the time.

Contract price

In practice, the actual quantities of work needed to be performed may vary from the quantities estimated by the Engineer and stated in the Bill of Quantities. The contract price is the total sum of money due to the Contractor for fulfilling the Contract. It is the amount obtained by applying the tender rates to the quantities of work actually performed, together with preliminary items, method-related charges, amounts for variations, contract price fluctuations, claims and all other amounts to which the Contractor is properly entitled under the provisions of the Contract.

Insurance

Under most conditions of contract, the Contractor is required to insure against specified liabilities, and the premiums for such insurances are included in the tender total. The greater the liabilities placed on the Contractor, the greater will be the tender total. The Promoter may insure against those liabilities which are his under the terms of the Contract, but some large organizations prefer to carry and finance the risk themselves. The Promoter should ensure that the Contractor's insurances provide the protection required.

Legal compliance

All work must be undertaken in accordance with the laws in force in the country where the work is performed. In addition the Contract Documents may refer to specific acts or regulations of special significance with which the Contractor must comply. It is the Contractor's responsibility to assess the significance of these and other legal obligations and to make appropriate allowances in his tender.

Qualified tenders

To enable the Engineer to compare tenders on a common basis, it is important that tenders are submitted without qualification or alteration to any of the Contract Documents. It is often stipulated in the instructions to tenderers that no qualifications will be permitted, but if it is not it should be made clear how qualified tenders will be dealt with in adjudication. Provided an unqualified tender is submitted, a tenderer should not be penalized for offering alternatives, so long as his intentions are properly described and full supporting information including pricing is submitted.

Alternative tenders

The instructions to tenderers should state whether offers based on alternative designs will be considered, and if so the procedure to be followed. Sometimes Promoters require any alternative proposal to be submitted in confidence before the end of the tender period so that the Engineer can give a preliminary view on whether or not it is likely to be acceptable.

Security for due performance by the Contractor

Most conditions of contract require the Contractor to undertake to provide a bond for the due performance of the Contract. This is usually provided by a bank or insurance company and entitles the Promoter to recover damages up to a prescribed amount (usually a percentage of the tender total) if the Contractor does not fulfil all his obligations under the Contract. The cost of such a bond adds to the contract price. It must be borne in mind that in most forms of contract the following intrinsic securities for the performance of contractual obligations already exist

(*a*) the Contractor's own plant which, under the *ICE Conditions of Contract*,[2] becomes the property of the Promoter when brought on to the site (it reverts to the Contractor on completion of its work when it is taken off the site)
(*b*) the retention money (generally a percentage of the measured value of the work up to a stated limit) held by the Promoter
(*c*) the materials supplied and work done by the Contractor during the period preceding payment by the Promoter.

Security for due performance by the Promoter

If a Promoter is of doubtful financial standing, a contractor may, before entering into a contract, require a bond or other form of security as a protection against default by the Promoter.

Time for completion

The time specified by the Promoter for completion of the whole or sections of the Works may affect the tender total. A requirement for speedy construction under threat of heavy damages for non-completion generally results in a higher tender. Speed in construction, whether on economic or political grounds, is best achieved by the Promoter's early decisions in ordering the start of the design work and avoiding delay in approving the details of the contract.

Cost of tendering

The cost of tendering is an appreciable part of a contractor's expenditure. Whether or not the tender is successful, this cost must be included in tender prices generally, and thus eventually must be borne by Promoters generally. Contractors sometimes tender when they do not wish to undertake the work in the belief that their failure to do so may prejudice future opportunities. When appropriate, the Engineer should invite such contractors to withdraw from the selected list, and make it clear that their failure to tender will not prejudice future invitations.

Selecting a contractor

Opening of tenders

Tenders should remain unopened and made secure until the appointed time for opening. At least two persons representing the Promoter should be present when the tenders are opened. They should record the totals of the Bills of Quantities and pass them to the Engineer for assessment. The Engineer should then report his findings to the Promoter and make appropriate recommendations.[11]

Covering letters

It is common for a tender to be accompanied by a covering letter, the contents of which may be important. In certain circumstances, a covering letter can be considered as part of the tender, in which case both parties should be aware that such matters as proposed methods of construction or proposed subcontractors which may be described in the letter become legally binding.

Scrutiny

The rates entered in the Bill of Quantities are the tendered rates. All tenders should be checked by the Engineer's organization for arithmetical accuracy by multiplying the quantity by the tendered rate. For the purposes of comparing tenders any error in the extension should be corrected. Errors in the summation of extensions should also be corrected to give the correct total of the priced Bill of Quantities. In order to preserve parity of tendering, correction of tender rates after submission of tenders should not normally be permitted. In these circumstances a contractor may be given the option of withdrawing his tender.

The Engineer should scrutinize the tender rates, and if he finds any which are excessively high or low, he should carefully consider the consequences on the outcome of the Contract, and advise the Promoter accordingly in his report. If apparent errors occur due to a misunderstanding of the Contract Documents and these are common to several tenders, it may be necessary to invite new tenders.

Pricing of a tender in a way which will result in high payments to the Contractor early in the contract period – sometimes called front loading – should be considered carefully, and the financial consequences recognized when tenders are compared.

Early notification to tenderers

When it becomes clear during scrutiny of the tenders that certain tenders will not be accepted, the tenderers concerned should be advised as soon as possible. The remaining tenders will be subject to closer examination but as soon as possible after the Promoter has accepted a tender, the unsuccessful tenderers should be notified.

Post-tender meetings and negotiations

Following scrutiny of the tenders, it may be desirable for the Engineer and Promoter to meet the lowest tenderer, or possibly the two or three lowest tenderers separately. Such meetings should serve to clarify any points of doubt or difficulty such as resources, methods of construction, organization, qualifications and alternative designs. The Engineer should not meet tenderers without the Promoter's prior agreement.

Lowest tender

The Engineer should not hesitate to recommend a tender other than the lowest if he concludes that it is in the best interest of the Promoter to do so. In making such a recommendation a clear explanation is essential so that the Promoter fully understands the consequences of his rejecting the Engineer's recommendation. In open tendering it may be necessary to reject a tender, either because of inadequate experience or because of doubtful financial standing of the tenderer. In selective tendering rejection of the lowest tenderer may be difficult to justify, especially in the public sector where existing rules or standing orders may specify that the lowest tender must be accepted.

Report on tenders

The Engineer's advice to the Promoter should take the form of a report setting out how he has scrutinized the tenders and giving his conclusions and recommendations regarding which tender to accept. The scope and detail of this report will vary according to the circumstances, but frequently the following are included

(*a*) a tabular statement of the salient features of the tenders received, e.g. tenderer's name, total of priced Bill of Quantities before correction of arithmetical errors, validity and qualifications
(*b*) arithmetical errors discovered and the effects of corrections on the bill total

(*c*) details of discussions held with any of the tenderers
(*d*) a concise summary of the analysis of each tender or, say, the three or four lowest tenders; reasons for considering any tender invalid, the consequences of any qualifications and discussion of any methods of construction proposed
(*e*) a tabular comparison of various sections of the Bill of Quantities and main rates with comments on unusual rates
(*f*) a comparison of tenders with the Engineer's estimate
(*g*) recommendations of the most acceptable tender
(*h*) recommendations for dealing with errors, qualifications and any points arising from discussion with the recommended tenderer, whether by amending the tender or by a counter-offer by the Promoter
(*i*) a financial statement indicating likely programming of payments to be made by the Promoter.

Acceptance of tender

Letter of intent

When it is not immediately possible to issue a formal letter of acceptance to the successful tenderer, it may be useful to issue a letter of intent to enter into a contract. Such a letter should be precise and contain at least

(*a*) a statement of intent to accept the tender at a future date
(*b*) instructions to proceed (or not to proceed) with ordering materials, letting subcontracts and so on
(*c*) a statement of what costs, if legitimately incurred, will be reimbursed to the Contractor if eventually no contract is made
(*d*) a limit to financial liability before formal acceptance
(*e*) a statement that formal acceptance will render provisions of the letter of intent void
(*f*) a request for acknowledgement of receipt and agreement to the conditions of the letter of intent.

Letter of acceptance

A letter of acceptance of the tender as finally modified and agreed between the Promoter and the selected contractor should be issued by the Promoter. This should inform the Contractor that all future correspondence regarding the administration of the Contract, other than any formal requirement to the contrary in the Contract, should be addressed to the Engineer. Alternatively, the Promoter may send to the selected contractor a counter-offer. This is an offer comprising the contractor's original tender but incorporating any modifications or amendments which the Promoter wishes to make.

The letter of acceptance or counter-offer may be issued by the

Engineer, on the written authority of the Promoter, provided that the existence of such authority is made known to the Contractor.

A legally binding contract is established when a tender is accepted by the Promoter or the Promoter's counter-offer is accepted by the Contractor. Formal acceptance in writing is not legally essential, but it is prudent and commercially preferable so as to prove the scope and existence of the agreement.

Notification of tender results

All tenderers should be notified of the tender results, including the amount of the successful bid, as soon as possible because their future tendering policy may be influenced by the results. This notification frequently consists of a list of tenderers and a separate list of tender totals thus enabling the tenderers to note their ranking without specific reference to the others.

Forms of agreement and bond

The Form of Agreement, if it is required by the Promoter, should be drawn up and executed by the Promoter and Contractor at this stage. If, in the rush to get work started, the agreement is left until later, difficulties may arise over the precise wording, particularly if the tender has been modified by agreement between the parties during the period between its submission and the issuing of the letter of acceptance.

The Performance Bond should also be executed by the Contractor at this stage, if it is required by the Promoter. In some cases the satisfactory completion of this document is a prerequisite to the issue of the letter of acceptance by the Promoter. A model Form of Agreement and a standard Performance Bond are given in the *ICE Conditions of Contract*.[2]

Overseas

Differences from the UK

The design and contract preparation for overseas projects are generally similar to those for comparable schemes in the UK, but certain modifications are usually necessary. The standard conditions of contract prepared by the Fédération Internationale des Ingénieurs-Conseils in association with certain other bodies (the FIDIC conditions)[12] are sometimes used as a basis for overseas contracts. Part II of the FIDIC conditions comprises conditions of particular application which are clauses which must be specially drafted to suit each particular contract. Many countries have their own standard conditions of contract which they require to be used in contracts prepared for them.

Law of the contract

The conditions of contract must state under which country's laws the Contract is to be construed. Normally this is the country in which the Works are to be constructed. These are likely to differ from English law, and care must be exercised when drafting Contract Documents to guard against possible conflict, especially in countries where the legal system is based on non-English traditions. Specialist legal advice may be necessary.

Language

The language to be used in the Contract Documents must be agreed between the Promoter and the Engineer. The language(s) to be used in correspondence between the Promoter, the Engineer and the Contractor should be similarly agreed and stated in the Contract Documents. When more than one language is used, the language according to which the Contract is to be construed and interpreted – the ruling language – must be stated. It is important that the Engineer's Representative, the Contractor's Agent and senior members of both staffs are fluent in the language used for the administration of the Contract.

Payment

Frequently, provision is made in overseas contracts for advance payments to the Contractor in respect of plant and materials, to assist the financing of the high cost of mobilization. It may also be necessary for some payments to the Contractor to be in foreign currency, in which case the various proportions should be clearly stated. Foreign currency payments should not be subject to variations in rates of exchange, but should be based on a rate of exchange prevailing at or some time before the date of tender, as stated in the tender documents.

Bill of Quantities and measurement

The method of measurement on which the Bill of Quantities is prepared should be stated in the Contract Documents or the basis of measurement should be stated in a preamble to the Bill of Quantities.

Where there is included with the Bill of Quantities or otherwise a schedule of basic prices (or rates) for use in the operation of a variation of price clause, the Contractor should ensure that all items to which he may wish the variation of price clause to apply are properly included, either at the tender stage or at the latest before a formal agreement is executed. He will not normally be able to add extra items at a later stage.

Construction

Initiation

Date for commencement

Shortly after the acceptance of the tender, the Engineer must notify the Contractor in writing of the date for commencement of the Works in order to initiate the execution of the project and to declare the date from which the time for completion will run. Thereafter the Contractor is responsible for proceeding with due diligence and completing the Works in accordance with the Contract.

Insurance

Apart from planning and executing the Works, the Contractor must ensure that the appropriate insurance cover called for under the Contract is arranged – in the joint names of the Promoter and the Contractor where applicable – and approved by the Promoter. The Contractor should remember that, except in stated situations, he is effectively indemnifying the Promoter from all damages and costs, and that, for instance, the level of third party liability cover required under the Contract is a minimum. The Contractor's risk is unlimited, and the minimum level of cover is not for him to decide. However, the extent of insurance cover not specifically required under the Contract is a matter for his decision.

Statutory notifications

The Contractor must give all statutory notifications of his new activity to the appropriate bodies, such as the Health and Safety Executive, but he may assume that all planning applications relative to the Works and specified Temporary Works have been obtained by or on behalf of the Promoter.

Finance

The Promoter sets his financial arrangements in motion in line with the cash flow projection of the Contract provided by the Engineer, who will keep him informed of the financial requirements of the Contract as it proceeds.

Programme and methods of construction

The Engineer is required to approve the programme of construction and the methods which the Contractor proposes to use, the details of which the Contractor must provide under the Contract if requested by the Engineer. Unacceptable proposals must be amended as necessary and approval obtained at the earliest possible date if work is not to be delayed.

Division of responsibility

On most contracts two independent but related systems of control and responsibility exist at the site: that of the Engineer and that of the Contractor.

Responsibilities of the Engineer

The Engineer

Where the Promoter and the Contractor have entered into a Contract for the execution of a project in the conventional manner, the Engineer must be nominated in the Contract Documents as 'the Engineer' and he exercises the powers reserved to him in that capacity for the administration and timely completion of the Contract. In carrying out certain functions he is also the agent of the Promoter, thereby acting in a dual capacity. In administering the Contract the Engineer's decisions must be scrupulously fair and impartial as between Promoter and Contractor, and must, to the extent that the Contract Documents provide, be based on the terms and conditions specified therein. Any restrictions imposed by the Promoter on the Engineer's authority to exercise his unfettered judgement or the full powers vested in him by the Contract Documents are undesirable and not in the Promoter's interest. Nevertheless, if any such restrictions exist they must be notified to the tenderers, otherwise the Promoter may be entering into the Contract under false premises. The position of the Engineer in relation to the Promoter's auditor should be made known.[1]

Specialist firms

The Engineer may need to employ, on behalf of the Promoter and at his expense, specialist firms for inspection and testing of the Works, and for taking samples of materials and testing, both on and off the site, as well as for the inspection and testing of fabricated work. Specialist firms may also become nominated subcontractors, if the Contract so provides, for both the detailed design, manufacture and installation of specialist work and/or plant items.

Delegation of authority

The conditions of contract may distinguish between those matters which must be settled by the Engineer and those which may be resolved by his representative on site. The latter is frequently called the Resident Engineer but more properly the Engineer's Representative, which is the name used in the *ICE Conditions of Contract*.[2] The extent to which the Engineer can delegate authority to the Engineer's Representative is laid down in the Contract. Within the limits so prescribed the Engineer may at his own discretion delegate authority to the Engineer's Representative. He should be influenced in this matter not only by the character of the work, but also by the capability and experience of the staff he employs at the site. The Engineer should inform the Contractor in writing of any delegation he makes of his powers to the Engineer's Representative. The Contractor should be given the right to appeal to the Engineer against decisions of the Engineer's Representative. It is good practice for the Engineer to notify the Contractor of the names and positions of responsible persons he will have to deal with during the execution of the Contract.

Direct labour

Where the Promoter has decided to carry out the project by direct labour, the civil engineer in charge, in addition to discharging all his engineering, administrative and co-ordinating duties, may be required to undertake many executive duties otherwise performed by a contractor. However, the Promoter is responsible for the financial risks involved in the execution of the Works, the provision of finance to meet the wages and salaries of the persons employed, the accounts for materials and plant required, and so on.[7]

Project engineer

The Engineer's head office is represented by a project engineer who may be a partner in the firm and whose duties are broadly

(*a*) to design in detail, amplifying the tender drawings by further drawings, which are instructions to the Contractor under the Contract
(*b*) to redesign if there are varied requirements of the Promoter or changed site conditions, and to estimate the effect which each of these variations will have on the programme and cost of the Works; these variations must be valued in accordance with the Contract
(*c*) to supervise the Engineer's Representative
(*d*) to prepare forecasts of expenditure and regular and special progress reports
(*e*) to check the monthly measurement of work done preparatory to the issue by the Engineer of interim payment certificates; it is desirable that agreement on payments remains as far as possible with the Engineer's Representative and the Agent

(*f*) to issue the Engineer's certificates for interim payment to the Contractor
(*g*) to settle claims and disputes and to issue completion and maintenance certificates.

Engineer's Representative

The function of the Engineer's Representative is to watch and supervise on a day to day basis the construction and maintenance of the project. Depending on the size of the project, the Engineer's Representative may have assistant staff under him. Such staff, particularly the clerks of works and inspectors, should be selected with regard to their practical experience of the type of work to be supervised. While his assistants deal mainly with detail, the Engineer's Representative must plan ahead and discuss future parts of the Works with the Agent of the Contractor to ensure that the phasing of the Works is properly planned to suit the approved programme. This close collaboration of Engineer's Representative and Agent also facilitates consideration of changes proposed by the Contractor, and the subsequent submission of such proposals to the Engineer for his approval. The principal duties of the Engineer's Representative are

(*a*) to organize his work to suit the approved programme
(*b*) to co-operate closely with the Contractor on matters of safety
(*c*) to supervise the Works to check that they are executed to correct line and level and that the materials and workmanship comply with the Specification
(*d*) to examine the methods proposed by the Contractor for the execution of the Works, the primary object being to ensure the safe and satisfactory execution of the permanent work
(*e*) to execute and/or supervise tests carried out on the site, and to inspect materials and manufacture at source where this is not done by the Engineer's head office staff
(*f*) to keep a diary constituting a detailed history of the work done and of all happenings at the site, and to submit periodic progress reports to the Engineer
(*g*) to measure in agreement with the Contractor's staff the quantities of work executed, and to check daywork and other accounts so that the interim and final payments due to the Contractor may be certified by the Engineer
(*h*) in the case of any work for which the Contractor may claim payment as additional work, to agree with the Contractor and record all relevant circumstances so as to ensure that agreement exists on matters of fact before any question of principle has to be decided by the Engineer

These reports must be in a form that will give the Engineer a clear and concise picture of the progress made

(*i*) to record the progress of the work in comparison with the programme

(*j*) to record on drawings the actual level and nature of all foundations, the character of the strata encountered in excavation and full details of any deviations from the Drawings which may have been made during the execution of the Works, i.e. to produce all record drawings.

Progress reports

The Engineer should maintain a check on progress through regular reports submitted to him by the Engineer's Representative. These reports must be in a form that will give the Engineer a clear and concise picture of the progress made and the extent to which this is ahead of or behind the Contractor's programme; they should be accompanied by such progress charts or diagrams as may be necessary for this purpose. If progress is behind programme on any items of work the report should state the reasons for the delay and the steps which the Contractor is taking to remedy matters. The Engineer's Representative may also be required to submit financial reports.

Instructions to the Contractor

All instructions should be given by the Engineer's Representative to the Agent in writing, either directly or as confirmation of verbal instructions. It is good practice to avoid giving instructions to other individuals within the Contractor's organization.

Variations

Any instructions from the Promoter concerning amendments to the design or to the work already done, on account of altered requirements, should invariably be given to the Engineer. The Engineer should then instruct the Contractor by means of a variation. The Engineer should also advise the Promoter on variations found necessary or desirable and inform the Promoter of the effect of all variations on the programme and the cost of the Works.

Knowledge of the Specification

Specifications need informed interpretation and some aspects may need to be varied and repriced accordingly. The resident staff should always refer any matter of doubt to the Engineer.

Inspectors' duties

Experienced inspectors should be employed under the Engineer's Representative to undertake general supervision of the Contractor's work, and junior inspectors as detail checkers on the mixing of concrete and any such work requiring constant supervision. The duties of inspectors are of great importance and demand wide experience, practical knowledge, integrity and tact in dealing with the foremen and workmen employed by the Contractor. Selection of suitable staff is therefore a matter of great importance not only to secure satisfactory work but also to ensure smooth working of the Contract. Inspectors must be above suspicion.

Value of co-operation

There is a great need for the Engineer and the Contractor to work in close co-operation and to understand the problems that the other has to face. Any advice, assistance and co-operation given to the Contractor in his execution of the Works is likely to benefit not only the Contractor but also the Engineer and the Promoter. Likewise any advice and assistance which the Contractor can give to the Engineer should benefit all parties. It is the common duty of the Agent and the Engineer's Representative to see that the Works are executed in accordance with the Specification and the Drawings. Experience, commonsense and judgement are required in the exercise of this duty, and in the exercise by the Engineer's Representative of the powers delegated to him by the Engineer under the Contract.

Responsibilities of the Contractor

Implementation of the Contract

The Contractor is responsible for constructing and maintaining the Works in accordance with the requirements of the Contract Documents. There should exist a strong bond of common interest between the Engineer and the Contractor because both should wish to see good construction materialize and want a successful outcome to crown their labours. However, neither must forget that the independence of the other should be respected; each is entitled to his own freedom of thought, outlook and need for privacy.

Freedom of the Contractor

It is in the best interest of all parties that the Contractor should be as free as possible under the terms of the Contract to execute the Works in the way he wishes. His preference for a particular design of and method of carrying out the Temporary Works and his idea of the order of sequence of construction may differ from those of others. However, he is better content, and therefore works better, when he uses his own ideas. Frequently there may be circumstances which necessitate restrictions. Part of the work may be wanted first for reasons outside engineering. In such cases the Engineer usually specifies the necessary requirements, but it may still be possible and desirable to allow the Contractor to submit alternative proposals for meeting them.

Subcontracts

Civil engineering contracts normally provide that the Contractor shall not place any subcontracts without the approval of the Engineer, and that the Contractor shall remain liable for all the acts and defaults of subcontractors. There is a need to distinguish between non-nominated or domestic subcontracts, where the Contractor selects the subcontractor for his own reasons, and nominated subcontracts, where the subcontractor is selected by others.

Nominated subcontracts

Where the nominated subcontract system is adopted, the Engineer obtains a tender from a selected firm for the supply of plant or materials or for the execution of special work, and includes in the Bill of Quantities a prime cost sum or provisional sum for the work in question. After the main contract has been placed the Contractor is instructed to accept the tender of the nominated subcontractor. As the Contractor then becomes liable for any default or financial failure of such a subcontractor, he should have the right to decline the employment of any nominated subcontractor against whom he has just cause for objection. It is therefore desirable that the Contractor should be called into the discussions on the employment of nominated subcontractors at an early stage.

The original tender of the nominated subcontractor should be based on the obligations and liabilities which the Contractor has towards the Employer under the terms of the main contract, otherwise the Contractor may be involved in financial and other liabilities which he could not have foreseen when tendering. The Contract usually provides that the Contractor may add a specified or tendered percentage to cover his responsibilities and services in connection with nominated subcontracts.

Design of the Temporary Works

The Contractor should submit drawings and design calculations for important Temporary Works to the Engineer, and the Engineer should scrutinize them with care. This check in no way relieves the Contractor of his responsibility for the adequacy of the design and construction of the Temporary Works. However, it does enable the Engineer more effectively to discharge his overall professional responsibility to the Promoter for the satisfactory execution of the project.

Such independent scrutiny provides valuable extra insurance against mistakes in the design of the Temporary Works. The Contractor should welcome it, and the Engineer should insist that it is done, in the interests of the safety of personnel and the Works. In the UK the operation of this practice has been made more complex by recent health and safety legislation.

Quality standards

More human lives depend on the safety of civil engineering structures than on the product of most other industries and quality standards must be jealously guarded throughout construction. At the planning stage the Contractor must make sure that the construction methods and plant employed by him can produce work of a quality not lower than the standard specified. On the site proper supervision and control are needed to ensure that output and quality do not come into conflict. To this end many of the larger contractors have set up extensive research and development facilities and quality assurance schemes which help greatly to combine increased output with high standards of quality.

Materials and workmanship

The Contract Documents specify in detail the quality of materials and workmanship required, and the tests which must be made regularly to ensure that the finished work complies with the Specification. The *ICE Conditions of Contract*[2] make provision in general terms for the execution of and payment for tests of materials and workmanship required by the Engineer.

Inspection by the Engineer

It is the Contractor's duty to ensure that every facility is given to the Engineer to enable him to inspect materials and manufacture at all stages both on and off the site.

Setting out

The responsibility for setting out the Works must be clearly defined in the Contract Documents. Under the *ICE Conditions of Contract*[2] it rests entirely with the Contractor, regardless of any check the Engineer's Representative may make. The complexity and accuracy of the survey methods employed in setting out depend on the site conditions and the type of work under construction. For example, survey methods giving a high degree of precision must be employed in the setting out of tunnels or for determining the span distances in bridge construction, whereas simpler methods will usually suffice for such work as locating roads in a new housing scheme.

The Contractor's organization

The Contractor's head office

It is not possible to trace more than a general pattern among contractors' head office organizations. Contractors include both public and private companies, and firms with varying financial and other resources and records of experience. Accordingly methods of business operation and details of internal organization differ considerably. The Contractor's head office is usually the centre from which the board of directors controls the whole of the organization. Engineers with experience in managerial control will often be in top executive positions and direct particular divisions or departments.

Division of the Contractor's organization

It is usual in the construction industry for the Contractor's organization to be divided into two main sections: technical and non-technical.

The responsibilities of the technical section include

(*a*) engineering services, including the preparation of designs for temporary and permanent structures and the planning and programming of work
(*b*) marketing, estimating and tendering, specifications, quantities, contract conditions and negotiations
(*c*) supervision of work production control and costing, progress reports, liaison with the Engineer, monthly interim and final valuations
(*d*) quality control, research and development, laboratory, site investigations and geotechnical processes
(*e*) central plant and transport depots with workshops and repair facilities, routine inspection of plant and equipment, purchase and sale of plant
(*f*) the training of technical staff.

Those of the non-technical section include

(*a*) secretarial and legal matters
(*b*) finance, accounts, audits, payments, cash and payroll checks
(*c*) insurance, licences, taxation and returns
(*d*) purchasing, expediting, shipping, carriage and invoices
(*e*) costing records and analyses
(*f*) plant and transport records and registers
(*g*) correspondence and records
(*h*) labour relations
(*i*) safety and welfare
(*j*) the training of non-technical staff.

Centralized servicing

In many large firms, particularly those with branch offices throughout the UK or overseas, certain activities (e.g. estimating, plant, transport and laboratory) are centralized.

Site staff

It is important for adequate and experienced staff both technical and clerical to be sent to the site at the start. If the Works are to proceed smoothly and economically much will depend on the prepara-

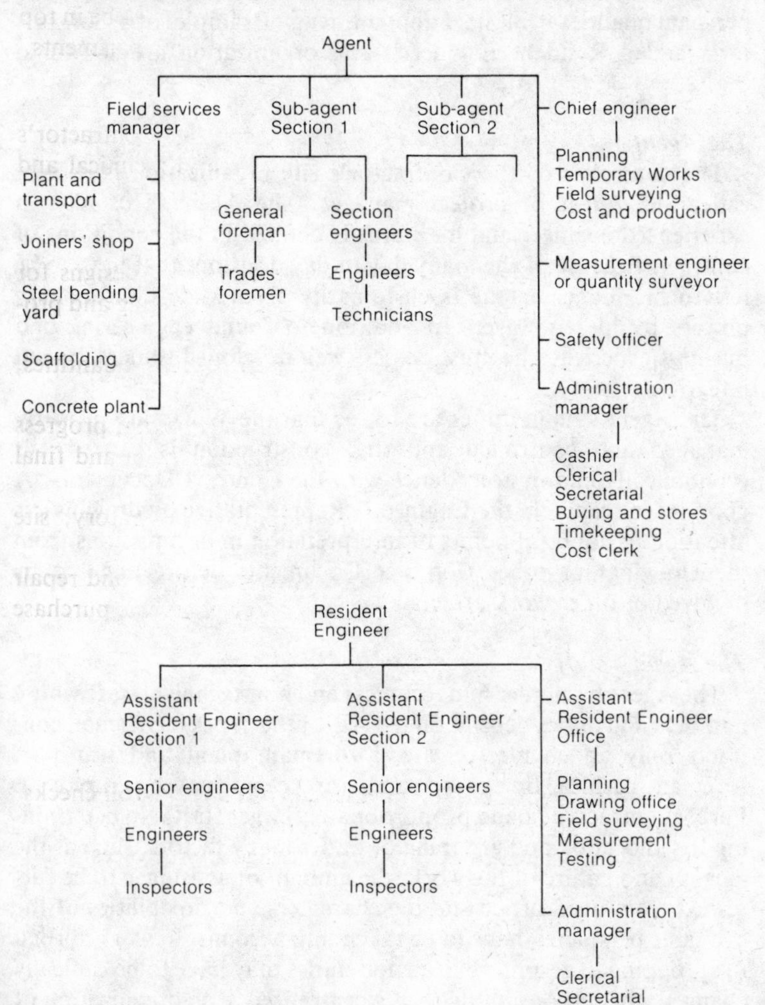

Fig. 2. Examples of a Contractor's and a Resident Engineer's site organization

tions that are made before labour arrives in any numbers. These preparations are likely to be inadequately planned if one member of the staff has to do several men's work during the initial period. Generally, a staff that has acquired the team spirit through previous employment in the same contracting firm will establish an efficient organization much more quickly than a staff lacking that common interest. For work in isolated locations the Contractor must take special care in the selection of site staff. Only men of proven ability and integrity should be considered and due regard should be paid to personal qualities in all staff appointments. Examples of a Contractor's and a Resident Engineer's site organization are shown in Fig. 2.

The Agent

The controller of the Contractor's site organization is usually called the Agent or project manager. The Agent is usually an experienced engineer and his status is defined in the conditions of contract. Because of the many day to day decisions that have to be left to his judgement he is customarily given wide discretionary powers by his employer. In addition to sound engineering and business experience he must possess well-developed leadership and integrity.

The Agent's main duties are to see that the Works are properly managed and controlled and that construction is carried out economically and in accordance with the Contract Documents. A good Agent can help the Engineer's Representative by drawing his attention to doubtful points of interpretation in or omissions from the Drawings or other Contract Documents before any cost is involved or other work affected.

The Agent's staff

The Agent has under him technical and non-technical staff whose numbers and duties depend on the size of the Works. On small contracts only an engineer, general foreman, plant and transport foreman, cashier, timekeeper and storekeeper may be employed. Larger contracts demand proportionately larger staffs. In determining the allocation and grouping of duties many factors, such as the locality and nature of the work, the amount of assistance to be furnished from head office and the characters and capabilities of the available personnel, have to be taken into account. In work abroad the grouping of responsibilities and duties may have to be radically changed, the responsibilities being centralized in an organization of highly experienced imported staff and the duties distributed to a large number of locally engaged personnel.

Chain of command

When allocating duties to personnel the aim should be to arrange the sequence of control and the routine flow of instructions and administrative returns to achieve simplicity consistent with adequate checks on all matters of importance. It is essential that every member of the staff knows from whom he must take instructions and to whom he may give them without undue interference. It is important also for the Engineer's Representative and his supporting staff to know with whom they should deal in the Contractor's organization. To this end a staff chart is usually prepared by the Agent in conjunction with executives from his head office. In the Contractor's site staff chart for a large project shown in Fig. 2, only one section of the job is shown in detail; the remaining parts are organized along similar lines.

The nomenclature is that normally used in the UK where the appointments of Agent and sub-agent are traditionally recognized as applying particularly to the executive branch, and senior engineer and assistant engineer to those rendering technical services. Supporting the Agent is a series of sub-agents and service managers. Also in Fig. 2 is a staff chart showing what the Resident Engineer's organization on the same job might be like. Normally the Contractor takes the lead in shaping his site organization and the Resident Engineer's organization mirrors, at least in broad terms, that of the Contractor.

Site technical staff

Sub-agents

The direct control of the various parts of the work is usually the responsibility of sub-agents. Depending on the size of the particular project they will have varying sizes of staff reporting to them but will usually have a number of section engineers responsible for the supervision of the actual operations. The control of labour and the operation of plant and transport are usually exercised through various grades of foreman according to the size and spread of the work concerned.

General foreman

The general foreman (who on larger sites is sometimes called the works manager) is an important link between the management and the foremen and gangers in direct charge of labour. His functions generally are to direct the day to day distribution of labour to particular operations under sectional or trade foremen, to supervise the flow of materials and stores and the disposition of plant to suit these operations, and to maintain and run all site communications. His office is an arduous one and his personal influence on the site organization can be a strong factor in achieving and maintaining efficiency. Instructions regarding labour and plant utilization on operations in hand and proposals in respect of their immediate future disposition should be discussed with and effected through the general foreman, if his usefulness is to be fully exploited.

Field services manager

The field services manager is responsible for the control of a number of services (Fig. 2). Typical of these are the concrete batching plant, the steel bending yard, and the plant and transport department. The sub-agents can call for these services as required on a daily or weekly basis.

Chief engineer

The chief engineer is responsible for the quality and accuracy of the Works through the section engineers. He has to check the coordination of the Drawings received from the Engineer and issue them to the appropriate personnel. He is responsible for any local designing that may be needed (especially of Temporary Works) as well as the general technical guidance of all staff.

Cost and production control engineer

The cost and production control engineer is responsible for keeping routine progress records and costs; he usually works through staff controlled by sub-agents. He is often responsible for routine measurement on site. An efficient costing system for controlling site operations requires work done to be measured regularly. It also requires information from time-sheets, plant cards, invoices and requisitions so that the costs of labour, plant, materials and equipment can be allocated to items of finished work and reviewed against budgets. Measurement of work done is necessary so that regular applications for payment can be prepared for submission to the Engineer. The Agent is thus able to compare regularly the costs he has incurred in doing work with its value in terms of the estimate and so to ascertain the efficiency and profitability of the Contract.

Progress and production planning are usually not the responsibility of the cost and production control engineer, but that of the chief engineer, because the planning of schemes of operation calls for wide experience and knowledge of constructional methods.

Section engineers

Section engineers are usually engineers with experience of both design and field work who, although ultimately responsible to the

Agent, report to the chief engineer on the quality, accuracy and control of the Works. Each section engineer must liaise with his foreman to plan the work of his section to be executed daily, weekly and monthly. He is also responsible for reporting matters of detail to the measurement engineer or the quantity surveyor. On large contracts where measurement is complex the team involved may be under the direction of a measurement engineer or quantity surveyor. However, the work of measuring, calculating and substantiating quantities for valuation purposes normally remains with the section engineers.

Measurement engineer or quantity surveyor

Apart from measuring and recording quantities of work done, there is the important task of ensuring that the Contractor is not only paid but paid as quickly as permitted under the Contract. This task is the responsibility of the measurement engineer or the quantity surveyor. He will prepare the interim and final measurements and valuations. He will request payment for additional or varied work. These applications have to be substantiated and agreed with the Engineers' Representative. If agreement is not achieved, the Contractor may give notice of claims, in which case it is the duty of the measurement engineer to keep records for their later submission.

The operatives

Labour relations

The civil engineering industry has a comparatively good record in industrial relations and every effort should be made to maintain it. The *Working rule agreement* of the Civil Engineering Construction Conciliation Board for Great Britain,[13] negotiated between the employers' federation and trade unions, contains the terms and conditions of employment and grievance procedures for workers in the civil engineering industry. A thorough knowledge of the agreement and the amendments made to it from time to time is a prerequisite of competent site management.

Communications and procedure

Good industrial relations on site will be difficult to maintain unless it is established that management is consistent, fair and reasonable. To this end a good communication system should operate and, within reason, management should always be prepared to receive the operatives' representatives and hear them with courtesy and patience. On sites where large numbers of men are employed, it is advisable to establish a procedure for anticipating and dealing with questions.

Labour unrest

Good management will sense impending labour unrest, investigate its possible causes, and take prompt action to resolve it before work is disrupted. The *Working rule agreement*[13] sets out procedures for referring disputes which cannot be resolved on site to negotiation off site. Final reference is to the Civil Engineering Construction Conciliation Board. The Board, which consists of representatives of employers and trade unions, will invariably reach a decision on the matter before it; both sides are expected to accept and adhere to the ruling. It is the right and duty of both the management and men to submit to the Board for adjudication such matters as cannot otherwise be resolved.

Incentives

It is customary, although not obligatory, in the civil engineering industry to use incentive schemes. The basis of a good incentive scheme is that it should give an operative of average ability the opportunity to earn a reasonable amount over his basic wage in return for increased production. Incentive schemes require both technical and psychological skill to formulate and apply, otherwise discontent can quickly arise.

It is advisable to establish procedures for dealing expeditiously with bonus queries. On some jobs circumstances can increase the bonus until it is such a large proportion of the basic rate of pay that it becomes in effect attendance money so that its incentive effect is largely lost.

Piece-work

In some types of tunnelling it has become traditional for workers to be paid piece-work rates. The incentive is based on the worker completing an agreed quantity of work before the end of the day or the week, as the case may be, and then being free to leave work. The incentive is therefore time off, rather than extra payment. This can be more effective than the bonus system.

Standard outputs

Because the combinations of operations in civil engineering differ from site to site, each contractor should devise his own scheme for each contract, so that it is suitable for the particular circumstances. Standard outputs have been compiled for bricklaying, shuttering, steel-fixing and many other operations. They apply to a simple environment where work can proceed with little hindrance, but they serve as a valuable guide when figures are fixed for the conditions of each site.

. . . employed mainly indoors

Weekly measurement

The setting of production targets and the bonus applicable to them should be discussed and agreed between the Contractor and representatives of the operatives, and thereafter altered only if circumstances justify changes. Weekly measurement of production and the calculation of bonuses require promptness, speed and accuracy.

Not all operations can be made the subject of a bonus by the direct measurement of output; operatives on such work should therefore be given a financial interest in work which attracts bonuses and so gain some benefit whenever bonuses are earned by their colleagues.

Site office administration

Administration manager

The administration manager is responsible to the Agent for the efficient administration of the non-technical side of the site organization. Although employed mainly indoors, he must keep in constant touch with outside matters connected with the duties of the staff under his control. He should prepare lists of duties and routine for the guidance of his clerical staff and he must be able to devise means of checking and counterchecking cash transactions and stock-taking, and be quick to detect waste or misappropriation. He must always maintain tactful contacts with the technical staff and assist them wherever possible. He must be familiar with the standing instructions of his head office and secure full compliance with their requirements regarding regularity and accuracy of accounts and routine items. He controls the payment of wages through timekeepers and cashiers as well as the purchasing and checking of the receipt of materials.

The administration manager's work also includes checking accounts, insurance, safety precautions, site welfare and general matters concerned with the operatives.

Timekeeper

The timekeeper should be well versed in the customs of labour employment and should know the operations on which labour is employed. Apart from recording the times of individual attendance and absenteeism and compiling pay sheets, he should, in conjunction with foremen and gangers, allocate these times to the different operations for costing purposes. Although he is responsible to the administration manager, he must frequently take instructions from the subagents and section engineers on the collection of labour employment information.

Purchasing

Some contractors negotiate and place all orders through their head office buyers. Others delegate the duty, in whole or part, either to regional offices controlling a number of contracts or directly to a particular site if it is large enough. Centralized purchasing with its established connections and ready access to suppliers' representatives has many advantages, e.g. in the purchase of goods involving careful negotiations in technical selection or terms of payment. Local purchasing may make it possible to use other sources of supply which are more economical and avoid delay.

Buyer

A buyer should have wide experience of the materials and stores in general use for constructional work, and of the usual trade designations of quality and price customs. He should also have some knowledge of commercial law, particularly that relating to the sale of goods. If his duties include negotiating subcontracts which involve the employment of labour at the site he must be conversant with the relevant employment legislation. He must work closely with the engineering staff, who generally determine the requirements to meet the programme. He must collaborate with the storekeeper in checking that goods will be delivered to the site in time, and in chasing deliveries when suppliers fail to comply with their obligations or where earlier despatch has to be negotiated.

Storekeeper

The storekeeper's duties, like those of the timekeeper, call for an understanding of outside operations and office routine in the recording of all transactions. They include the checking in and safeguarding of all materials received. In this the storekeeper may have to work in conjunction with an engineer to check any special deliveries and with the general foreman in arranging for the correct distribution of goods on the site as they are delivered. Similarly, he must check in all stores and tools received and issue these for consumption or loan on the Works under a system of authorization laid down by the management. The usefulness of the store is judged by its ability to serve the immediate needs of construction, and its efficiency in the avoidance of losses or deterioration. The avoidance of extravagant issue or shortage of stock is generally under the control of the administration manager. On some projects the storekeeper is also responsible for maintaining the plant register, dealing with transfers from and to other sites, keeping records of tool sharpening, salvage recoveries, stock sales and other matters required for costing or accounts.

Accounts staff

The duties of an accounts or invoice clerk in a site organization will depend largely on the accountancy system used at the Contractor's head office. Accountancy requirements may vary according to the contractual or technical conditions of each job or to the ability of the personnel employed. Thus the head office system should be sufficiently flexible to adjust its demands for site returns to the prevailing circumstances. The ultimate purpose is to enable checked particulars of expenditure and receipts to be readily available for entry under defined headings in the Contractor's main account books at his head office.

Accounts system

An accounts system should be capable of

(*a*) stating expenditure to date, commitments and liabilities incurred, in a form useful to the Agent
(*b*) summarizing relevant information to aid the management in assessing the present and future financial state of the Contract
(*c*) providing up to date accurate financial statements at regular intervals and realistic estimates at interim periods
(*d*) preparing company accounts and returns in conformity with statutory requirements.

Site accounting

Any accountancy method will have to be divided into those duties performed at the site and those carried out at head office. The site duties will normally include

(*a*) accounting for cash payments of wages and site expenses
(*b*) estimating weekly cash requirements in advance
(*c*) checking and coding invoices against all deliveries.

Insurance

The subject of insurance is a specialized one, the extent of cover desirable and obtainable depending in each case on the conditions of contract and the particular risks involved. The matter can best be dealt with at the Contractor's head office where direct contact with insurance brokers experienced in this class of work can be readily maintained.

Site services, plant and transport

Site services

On site many services ancillary to the physical construction are required. Particularly important are plant and transport.

Plant and transport managers

Usually the types of plant and machine required for construction are determined by the Contractor's head office. The running and maintenance on the site is supervised by plant and transport managers who are supported by foremen and mechanics, preferably with experience of field conditions. On large and specialized works it may be necessary to employ fully qualified mechanical and electrical engineers. The uninterrupted operation of plant and transport depends on good site maintenance and this, in turn, demands well-equipped workshops, ample availability of spare parts and efficient storekeeping.

Maintenance of plant

Maintenance of plant falls into two categories: regular overhauls and day to day servicing. Overhaul consists of the stripping down of either the whole or a section of a machine and its reconditioning by replacement of parts which are so worn or fatigued that they will not last until the next overhaul. Servicing consists of daily and weekly inspection and action to ensure that everything possible is done to keep the machine working. It includes cleaning, lubrication, water checks, the tightening of bolts, adjustments, minor renewals and the taking of precautions against frost. A standardized routine for such periodic inspections is very important.

The object of maintenance is to avoid delay due to breakdowns. All delays should be recorded and the causes investigated so as to discover whether the breakdowns are due to defective design, materials or manufacture, or to bad maintenance or misuse.

Plant records

For every item of plant, records should show, besides the amount of fuel consumed and the hours of work, that the daily or weekly servicing has been properly carried out and, in the case of overhauls, details of replacements. Such records form not only a valuable history of each machine, but also the statistical basis from which are predicted the working times that the various parts of the machine will last before renewal. The history sheet will also show the information needed for deciding when a machine should be scrapped because it is no longer economical to run, and will assist in determining, or amending, the frequency of overhaul of such machines.

Construction planning and control

Scheme of operations

In order to carry out work efficiently a scheme of operations should first be decided on by those directly responsible for its execution. Decisions regarding methods, Temporary Works, plant and the like, and requirements of labour, materials and transport, all have to be made at the start of a contract. The layout of temporary buildings, offices, stores, workshops and temporary roads and railways needs careful consideration, because the location of these features in relation to the Works will affect the convenience and economy of future construction and administration.

Planning

Making decisions about how and when work will be done is called planning. There is often a tendency at the beginning of construction works, before any properly studied sequence of operations exists, to hurry labour, plant and materials on to site in order to make an early start. This usually results in confusion, subsequent delay and increased cost. The programme is an important document, and should be discussed and agreed by all concerned before work starts.

Master programme

Outline programmes will have been prepared by the Engineer and the tenderer. The Engineer's programme will have been included with his report to the Promoter; the tenderer's may have been submitted with or subsequent to his tender. Neither of these programmes is prepared in sufficient detail for the actual execution of the work.

One of the first duties of the Contractor is to submit a master programme for the Engineer's approval. This is an important control document and should show the periods during which the individual sections of the Works are to be carried out so as to ensure the correct sequence of operations and the completion of the Works within the time specified. All those in authority on the work – the Engineer's Representative's staff and the Contractor's staff – should be thoroughly familiar with this document and should constantly strive to carry out the work as planned.

If subcontractors are to be employed they should be notified well in advance and agreement reached so that they can, with proper planning and suitable mobilization, perform their work properly in the time allowed. The programme should provide the Engineer with dates by which detailed drawings will be required, and the Engineer and the Promoter with the dates when various sections of the Works will be completed and ready for use or for the installation of plant by other contractors. It should always provide essential information in regard to material, labour and plant requirements. The master programme also helps to discourage undue changes in design and layout.

Detailed programmes

Each section of the project should have its detailed programme dovetailed into the master programme.

Methods of programming

The most widely used forms of programme are bar charts and network diagrams. Bar charts show the programme in a form that can be easily read by the user, and provision is made for comparing eventual progress. Network diagrams have the advantage of showing the sequence and interdependence of activities and of clearly indicating the effects of delays on the programme. Networks may be drawn as an arrow diagram or a precedence network. On large and complex jobs they are normally analysed by computer so the effects of changes in phasing or resources can be forecast quickly and economically. Computer-aided network programming is complex and has tended to become the province of experts. There is a large amount of literature on the subject including two British Standards.[14,15] However, most run of the mill jobs can be programmed satisfactorily by using simple precedence diagrams and/or bar charts which can be prepared and updated without computer help. A simple precedence network and the bar charts prepared from it are shown in Appendix 4.

Modifications to programme

The formulation of a programme at the outset does not mean that it cannot be changed. A criterion of a good programme is that it shall be flexible enough to permit modifications to meet likely contingencies. This apparent paradox has sometimes been used as an argument against the making of programmes, on the grounds that, although one of the arts of construction is to reduce the unknowns to a minimum, the unexpected occurs so frequently that any planned sequence of operations is disturbed. Experience shows that it is exactly in such circumstances that the programme is of value, for it enables those in charge of the work to see what the effect of the contingency will be on subsequent operations and to adjust their plans accordingly. The working programme should therefore be continually updated. The master programme, however, may have contractual status and cannot be altered at will. Thus on one job it is possible to find two different programmes: a contract programme and a working programme.

Progressing

Taking systematic steps to see that the programme is followed as closely as possible is called progressing. Without this the value of the programme is largely lost. In order that progress may be compared properly with the programme a standard of measurement must be applied to each item. The measurement may be rough or detailed; it may be by volume, area, length, weight, number or other physical dimension. Measurement by time alone, or by man-hours, cost or money expended is meaningless because it gives no real indication of what has been physically constructed. The rule, therefore, is to measure by quantities of completed work. If at the outset these are not known for all the operations, they must be estimated, however roughly, so that every operation can be included. This rule applies equally to preliminary and temporary works, all of which should be measured (Appendix 4, Fig. 5).

Materials planning

The programme allows the buyers to draw up a schedule of materials required, to purchase accordingly and to ensure timely deliveries.

Labour planning

Similarly, the programme provides a forecast of labour requirements. Tender prices are usually based on labour outputs for each operation. Using these and the rate at which the operation is planned, the total estimated labour for the operation can be found. A chart of labour requirements can then be produced, showing by categories (skilled trades and labourers) the total number of men expected to be required at any particular time. The drawing up of this labour chart at an early date may show that there will be uneconomic peaks in the numbers of particular categories of labour required. By planning labour requirements early in a contract these peaks can be avoided by minor adjustments to the programme.

Plant planning

The amount of plant that will be required and the periods during which it will be employed must be determined as accurately and as early as possible, to enable decisions to be taken as to what plant will be available from other contracts, what new plant should be purchased and what hired. These are important points because, although in countries like the UK it is generally economical to do everything possible by machine, heavy plant is expensive and new plant represents considerable capital outlay. The rate at which each operation is to be carried out will indicate the number of machines required, and consequently the total number of machines of each type needed on the work at any one time. The programme provides the means of obtaining this information.

Use of plant

To derive the best value for money expended on plant, it is essential first to choose the correct machines for the task in hand, and second to arrange the execution of the work so that the machines are used to full capacity. To ensure the latter, it is important to remember that production that uses expensive plant must not be held up for the want of adequate transport of materials to or from the plant, or by the want of sufficient manpower. It is advisable when much plant is involved to draw up a plant operational chart covering the whole work. Such a chart should be similar to the programme and progress chart, and should show for each item of plant the sequence of operations on which it is to be employed. Such a chart also enables allowances to be made ahead of time for periodic machine overhauls.

Programming the plant to be used on any job can only be done by planning the output for each machine. As the work proceeds this must be watched and, for each machine, records kept showing its output, delays incurred and their causes. Such records also form the basis of statistics for planning future work.

Cash flow control

High interest rates mean that great attention must be paid to cash flow. At the tendering stage a complete cash flow analysis will be required in order to ascertain the amount of money employed at various stages of the work and the rate of cash recovery. During the execution of the work it is important that the prediction is monitored closely so that the actual cash flow position is known and can be compared with the estimate.

Costing

Costing is the calculation at frequent intervals of the costs of units of production to indicate to management whether or not the job is being run efficiently. This is to be distinguished from financial accounting, which is concerned primarily with the historical recording of receipts and payments of cash, amounts owed and the preparation of company accounts required by law.

Cost clerk

An experienced cost clerk can provide the information necessary for the financial management of the Contract. Through systematic analysis of pay-rolls and stores returns he can also assist the administration manager in tracing irregularities in timekeeping and store issues.

Costing system

A site costing system on civil engineering works is an essential part of a good works organization. It should have three objects

(*a*) to provide accurate unit costs at regular intervals of all the principal operations and related site overhead charges
(*b*) to derive cost data for partially completed works on which estimates of future liabilities for completion can be based
(*c*) to provide cost data for completed works to assist in future tendering.

Accuracy of cost items

It is important to realize that no matter how elaborate and efficient the costing system may be, its accuracy is largely dependent on the basic information provided by foremen, gangers, machine drivers and other operators. The cost items should therefore be physically recognizable on the site by these men and the cost clerks who record the information. The items of work listed in the Bill of Quantities are not necessarily suitable for costing purposes and rearrangement may be necessary.

Costing of labour and plant

Labour and plant are the key items in a costing system because they can go badly astray if they lack proper control. The entire labour force should come within the costing system and the total cost in any statement should agree with the total wages paid. Similarly the costing of construction plant should include all the hours which the plant spends on the various operations, whether it is active or not. Labour and plant costs should be produced weekly.

Costing of materials

With materials there is less likelihood of serious loss and, provided reasonable precautions are taken against gross wastage, there is no need for the costing system to produce material costs as frequently as labour and plant costs. A check on the usage of materials can usually be made monthly at the same time as the Works are measured for the monthly progress payments.

Costing details and allocations

The details to be included in the costing system should be decided by the cost and production control engineer, and all doubtful allocations should be referred to him. Data on costs may be useless for comparative purposes and misleading as guides in future tendering if their compilation is treated as a mere clerical duty without proper regard for the technical conditions in each case. So that costs may be used safely at a future date by an experienced estimator, they should be accompanied by adequate description of all the environmental conditions.

Safety

Background

The safety of all personnel working on building and civil engineering construction sites is a major responsibility of those involved, whether they be management or operatives. The construction industry is one of the most dangerous in the UK, accounting for nearly 20% of all serious accidents and 40% of all fatalities at work. Neither the industry nor the country can afford the enormous cost of loss of production which accidents cause. The number of man-days lost due to industrial accidents far exceeds that lost due to industrial disputes. Most of these accidents could be prevented. Forethought and planning are essential at the start of any project to identify the hazards and minimize their dangers.

A knowledge of, and dedication to, accident prevention are integral to good management and essential equally to the Engineer and the Contractor. Acts of Parliament and regulations play a vital part but ultimately it is up to the Engineer and the Contractor to recognize and demonstrate that safety and economy can be partners rather than rivals. The industry has for some time been keenly aware of the situation. Employers' federations, trade unions and the Institution of Civil Engineers have all been active in promoting the need for safe working practices. The Institution has organized conferences and published literature on safety on construction sites.[16–19] In addition, the Federation of Civil Engineering Contractors has produced a safety booklet for supervisors[20] and runs accident prevention courses for supervisory staff. Reference should also be made to *Safe work in sewers and at sewage works*.[21]

Co-ordination

Safety on site is the responsibility of the Promoter, the Engineer, the Contractor and all employed there. Representatives of all parties involved should meet to consider safety aspects in close co-operation.

Legislation

Numerous acts of Parliament and regulations may be applied to the construction industry in the UK (see Appendix 5). It is the responsibility of the Health and Safety Executive to enforce them. If the advice of a qualified safety officer is not available, use should be

. . . accident prevention . . . integral to good management

made of guidance literature published by the Health and Safety Executive and the Federation of Civil Engineering Contractors, and also the reference books listed in Appendix 5 which state the law and comment on its application in practice. Similar sources of information and advice should be used in planning and constructing projects in other countries.

Welfare

Welfare officer

On large contracts, especially those where a construction camp is provided, a welfare officer should be employed to see that living conditions and food are in keeping with the requirements of the local health authorities. He must listen patiently to all complaints – real and imaginary – and do everything reasonable to remedy them. He must also arrange recreational facilities and entertainment. Such duties are particularly onerous and call for wide experience in dealing with all types of labour. A camp superintendent should be employed to look after the day to day running of camps and eating arrangements.

Medical services

On the medical side responsibility rests with the resident or visiting medical officer to whom the nursing and orderly staff report. Casualty arrangements depend on the number of men employed and on the proximity of the site to public hospitals. For the treatment of accidents on site, men trained in first aid should be located at convenient centres throughout the work. Where diving and work in compressed air are involved, it is a statutory obligation to appoint a doctor and notify a suitable public hospital of the work, but if the nearest casualty hospital is too distant for the speed of service needed in an emergency medical facilities must be provided at the site.

Legislation requires special medical services to be available for hazardous work such as that in compressed air, radiation zones and diving (see Appendix 5).

Contract records

Every document generated by the execution of the Works – be it a letter, a minute of a meeting, a monthly measurement or a purchase order – becomes part of the contract records. Records whose usefulness is not wholly clear at the time but may become apparent later can help enormously with such things as accidents, failure or deterioration of completed work, an unforeseen need to do new work adjoining completed work (especially if such work is buried), disagreements between the Engineer and the Contractor over payments or delays, and designing and pricing future work.

Records needed later include

(*a*) diaries – every engineer on a contract should keep a detailed diary of his own work, however futile this may seem at the time
(*b*) private notes of verbal instructions or agreements that have not been confirmed in writing
(*c*) superseded drawings, which must be kept because something done when the drawing was current may two years (and five revisions) later seem very odd
(*d*) as-built drawings – everybody agrees these are vital but in practice the need to break into old work often reveals that the record drawings are incomplete or lost.

Payment to the Contractor

Monthly statements for interim certificates

A monthly statement showing the amount earned during the month in question is prepared and submitted by the Contractor in accordance with the Contract. The precise form of this statement is usually laid down by the Engineer. It shows the total quantity of work done to date under each item of the Bill of Quantities in accordance with the monthly measurement previously made and agreed between the Agent and the Engineer's Representative, with appropriate additions for additional works ordered and constructed to date and for the value of materials on site, and with appropriate deductions for retention money and amounts previously paid. The statement should be forwarded by the Engineer's Representative to the Engineer who will, after verification, prepare and send to the Promoter a certificate showing the amount payable to the Contractor. The Promoter is under a contractual obligation to make payment to the Contractor within a specified period.

Variations and additions

Variations and additions to an admeasurement contract are in one of the following categories

(*a*) variations or additions ordered by the Engineer (other than an increase or decrease in the quantities of work to be done under items in the Bill of Quantities) that will affect the cost of the Works and are necessitated by changed site conditions or alteration of the Promoter's requirements
(*b*) additional payments falling due to the Contractor under a contract price fluctuations clause where such a clause is included in the Contract.

A monthly statement . . . is prepared and submitted by the Contractor

Variation orders

Variations and additions ordered by the Engineer should be the subject of variation orders. Variation orders should be serially numbered for each contract, and should specify in full detail the additional work required to be executed and the Works which, although included in the Contract, are to be omitted as a result of the variations. Variation orders should be prepared and signed by either the Engineer of his representative and the prices for varied or additional work assessed as laid down in the Contract.

Requests by the Contractor

Variations requested by the Contractor may be considered by the Engineer or his authorized representative. If the Engineer decides to adopt them, he takes responsibility for them and treats them as variations ordered under the Contract. Changes not within the Engineer's authority should be referred to the Promoter.

Contract price fluctuations

Where a contract price fluctuations clause is included in the Contract, sets of indices produced from official records of the increased cost of commodities, labour, fuel and so on (such as the Baxter and Osborne indices in the UK) should be used to ascertain the payments due on each month's certificate. If such a system is not used the Contractor should be required to produce to the Engineer's Representative vouchers in support of any claim made under the clause.

Disputes and claims

The procedure to be adopted in the settlement of disputes that may arise during a contract is laid down in the Conditions of Contract. In most contracts a matter does not contractually become a dispute until the Engineer has formally given his decision on it and one or other party to the Contract (i.e. the Promoter or the Contractor) has formally objected to that decision. It may then be settled by negotiation between the parties (preferably conducted through the Engineer) or, if this fails, by reference to arbitration.

The most common kind of dispute, where the Contractor has objected to a decision of the Engineer, is usually termed a claim. The Contractor claims payment for expenditure he has incurred, or loss he considers he has suffered, on account of circumstances which he considers he could not have been expected to allow for in his tender; the Engineer must decide whether or not there is provision in the Contract under which he can properly certify this expenditure for payment by the Promoter.

Recording of facts relevant to claims

It is the duty of the Engineer's Representative and the Agent to agree and record full details of all the facts and circumstances relevant to any matter that may be the subject of a claim. Where agreement on facts cannot be reached, separate records should be kept and the reasons for disagreement carefully noted. Such records are essential to both the Engineer and the Contractor for the subsequent adjudication and settlement of claims.

Assessment of claims

The assessment of a claim is a matter for the Engineer after discussion between him and the Contractor. The Engineer should consider all relevant facts and circumstances in assessing claims. Having ascertained all the pertinent facts the Engineer will decide to what extent he considers a claim is justified in principle. Having decided this he should use factual data compiled from the records to price the claim and report his decision to the Promoter and the Contractor.

Completion certificates

When, in his opinion, the Works have been substantially completed, the Engineer should issue a certificate to that effect, and the maintenance period then normally begins. The Engineer may also at any time give such a certificate for any completed part of the Works and he must, if requested by the Contractor, give him such a certificate with respect to any substantial part of the Works which has been both completed to his satisfaction and occupied or used by the Promoter, or anyone acting on his behalf or under his authority.

Retention money

Generally the Contract stipulates that a percentage of the value of work completed and other items as measured monthly for payment to the Contractor shall be retained by the Promoter during the period of construction. Part (usually half) of the retention money becomes due when the completion certificate is issued, and the balance at the expiration of the period of maintenance.

Maintenance of the Works

The Works must be delivered up to the Promoter at the end of the maintenance period in the condition required by the Contract (fair wear and tear excepted) and the Contractor must complete any outstanding work and also make good any defects during the maintenance period or immediately thereafter. If any items of maintenance for which the Contractor is responsible remain to be carried out after this period, the Promoter is entitled to withhold from the balance of the retention money the estimated cost of such work until the Contractor has completed it. Failing this the Promoter may complete it himself at the Contractor's expense.

Maintenance certificate

The Works are not considered as completed until a maintenance certificate has been issued by the Engineer which states the date on which the Works were completed and maintained to his satisfaction.

Final payment

It is good practice to assess the final balance of payment due to the Contractor under the terms of the Contract before the end of the period of maintenance, so that when the maintenance certificate is issued it is possible to make the final payment to the Contractor. However, if at this time the Contractor still has outstanding claims, these should be settled under the procedure stated in the Contract.

Arbitration

If on account of failure to settle a claim, or for any other reason, a dispute still remains between the parties in connection with the Works it may be referred to arbitration under the conditions prescribed in the Contract.[22,23] Arbitration should be resorted to only after ultimate failure of the parties to agree by negotiation. The procedure leading to arbitration is explained in Appendix 6.

Overseas

Welfare

In overseas work the welfare requirement is even more important and becomes a bigger task than for work in the UK. Whether or not staff are accompanied by their families, full consideration should be given to the availability of medical and dentistry services, recreational facilities, leave arrangements and compassionate leave and to the difficulties of living in unfamiliar surroundings, in addition to the normal welfare and medical requirements for projects in the UK.

Offshore procurement

In the case of work abroad, special expertise is required in placing and arranging for inspection of export orders for constructional purposes and in dealing with shipping agents. The ultimate responsibilities regarding suitability and quality, where acceptance by the Engineer is given only on site, should be carefully defined.

Training and management

Training

Initial training

Set standards are required by the Institution of Civil Engineers for the training of chartered civil engineers, and by the Society of Civil Engineering Technicians for the training of technicians and technician engineers. The achievement of them depends on senior engineers in the industry offering new entrants training and experience, not only in the practical methods of design and construction, but also in the handling of plant and labour.

In-career development

It is the professional duty of civil engineers to keep themselves up to date with technical and other developments. They should do so by private study, attending engineering meetings, in-company training and by taking courses run by the Institution, universities and colleges. Engineers progressing towards top-level management will benefit from a mid-career business management course.

Industrial training boards

In training matters the advice resources and skills of the industrial training boards should be used. The board principally involved with civil engineering is the Construction Industry Training Board but also concerned are the Engineering Industry Training Board and the Offshore Industry Training Board.

Management

The engineer and management

The young civil engineer needs an understanding of how costs, programmes and contract decisions may affect his work. Management is therefore not something remote to be encountered later when his career is well advanced. The engineer may be involved in these and other managerial problems early in his career. As soon as he is given responsibility for organizing the work of a few assistants his career as a manager has started and he must think ahead. Technical competence is then no longer likely to be sufficient. Short courses, study and learning from experience in leading people, planning, budgeting, contracts and industrial relations may well be

Good communications are vital at all stages of a project

needed. The joint institutions' diploma in engineering management of the Institution of Mechanical Engineers and the Institution of Civil Engineers should be taken so that the engineer can test and demonstrate that he has acquired the relevant expertise.

Co-ordination

More demands on the engineer are made when further promotion brings responsibility not only for subordinate engineers but also for non-engineers. The latter may include accountants, purchasing officers, labour officers, camp superintendents, foremen and clerical staff of various grades. At this level, the civil engineer must learn to apply his trained mind to the understanding and appraisal of a wider variety of activities. In particular, he must develop the ability to see those activities as a whole so that he can co-ordinate them effectively. He must acquire the skill and ability to negotiate with people outside his own organization, e.g. promoters, contractors, staff representatives and trade union officials. Whether his ultimate choice of career be in design, research, teaching or construction, the same considerations apply. Career progression inevitably entails the organization of the work of subordinates, more forward planning and the methodical running of an efficient organization.

Planning

In the broadest terms, management consists of planning, organizing, staffing and controlling. On its highest level, planning involves the determination of a long-term policy for some desired purpose, and making the financial, legal and other provisions to prepare the way. At every subsidiary level of responsibility planning is needed.

Staffing

Most engineers need to be able to plan how to use the expertise and talents of varieties of people in a way which motivates them to be efficient and innovative. To achieve this the engineer needs to learn how to organize, define responsibilities, delegate authority and cultivate leadership.

Communications

Good communications are vital at all stages of a project – not only between the Promoter, the Engineer and the Contractor, but also between individuals within their own organizations. There is a special managerial skill in establishing and maintaining a balance between formal communications and the informal consultation and collaboration at all levels which are so important for the working of any organization.

Management literature

There is an extensive literature on all aspects of management, but the young engineer is recommended to avoid the detail to be found in the larger treatises and to confine himself to the principles. Early in his career he should seek to understand people as well as materials and develop that flexibility of mind which will enable him to adjust his outlook as his duties change with promotion. For this purpose he can study the civil engineering management guides being published by Thomas Telford Limited. Later he should read widely on such subjects as economics, trade unionism, human relations, the growth of industrial wealth and the history of ideas. From this wide range of interests, together with experience of works at home and overseas, there grow two indispensable qualities for high managerial responsibility: maturity of judgement and clarity of vision.

Appendix 1

Check-list of some of the duties of the Engineer

As a result of the criticism that engineers are not acting independently when considering contract matters which involve the Employer's and the Contractor's interests, the ICE Advisory Committee on Contract Administration and Law has emphasized the different roles the Engineer must assume under the fifth edition of the *ICE Conditions of Contract*[2] in pre-contract matters where he is the Employer's agent and professional advisor and, once the Contract has come into existence by the acceptance of a tender, his additional and quite separate duty to act impartially in the administration of the Contract in a quasi-judicial role between the Employer and the Contractor.

The following check-list will remind engineers of their duties and obligations under the *ICE Conditions of Contract*. The list should not be taken as a set of rules, but each item should be carried out where appropriate. The committee strongly recommends that the Engineer should be a named individual.

Pre-Contract duties

1. Ensure that the Employer is aware that he carries the financial risk for unforeseen events and of the financial managerial and advisory resources required for the Contract.
2. Warn the Employer of the decisions and actions required of him, giving programme dates for the finalization of designs, the provision of access, construction and taking over of the Works.
3. Design and detail the Works and as far as possible prepare clear working drawings and a concise Specification.
4. Prepare accurate Bills of Quantities, detailing the works required and complying with the Standard Method of Measurement where possible. Keep provisional items to a minimum (see the *ICE Conditions of Contract*, clause 57).
5. Ensure that the Employer and his staff understand the role of the Engineer under the *ICE Conditions of Contract* in ensuring fair dealings between the Contractor and the Employer.[24]
6. Adopt the *ICE Conditions of Contract* (or, for ancillary works, national and well-understood conditions of contract) in full, without any variations or deletions, and draw the Employer's

attention to the powers and duties of the Engineer under the Contract.

7. Ensure that the Employer and his Auditors accept the ICE/CIPFA joint statement on the relationship between Engineers and Auditors[1] and accept that the Engineer has the quasi-judicial powers to make decisions that are final and binding on the Employer and Contractor, subject only to reference to arbitration.
8. Ensure that the Engineer has a defined and readily understood method of selecting tenderers and recommend that the number invited should be limited.[25]
9. Ensure that all tenderers receive the same tendering information and are given a sufficient period for the preparation of tenders.
10. Make all site and service information in the Employer's and Engineer's possession available to those invited to tender.
11. Ensure that tenders are delivered in specifically marked envelopes to the Employer or the Engineer by a fixed date and time, and that they are opened in the presence of witnesses at a declared fixed time.
12. Check tenders carefully and correct any errors in the extension of items, rates, times and quantity. Notify tenderers of any resulting changes in the totals of the priced Bills of Quantities and tender sums.

    Review tenders received with particular regard to the proposed construction methods; the degree of risk involved; the implications of sectional completion dates on the Employer's and the Contractor's cash flow; and the anticipated final contract price. Submit a report to the Employer pointing out any rate that is less than the known cost of carrying out the work to which it refers and recommending the acceptance of a particular tender, with reasons. If any of the tender rates are in doubt, recommend that the tenderer be invited to stand by his rates or to withdraw.
13. Advise the Employer to give tenderers the name of the successful tenderer at the earliest opportunity and recommend that the list of the values of the tenders received be circulated.

Post-Contract duties

14. On the appointment of the Contractor, confirm the appointment by letter.
15. State by letters to the Contractor and Engineer's Representative details of the powers and responsibilities which are to be delegated to the latter. Name the Engineer's Representative and the members of the project team, and give a date for the com-

mencement of the Works (see the *ICE Conditions of Contract*, clauses 2 and 41).

16. Agree the extent and methods of payment for variations and extras. Agree methods for the supervision and recording of dayworks, preferably before work is commenced, and confirm these details in writing (see the *ICE Conditions of Contract*, clause 51).
17. Do not exceed the powers granted by the Employer, e.g. do not take on responsibility for re-design or significant variations and extra works without the Employer's agreement to such works and to the provision of the necessary finance (see the *ICE Conditions of Contract*, clauses 17 and 44).
18. Make decisions on extensions of time at the stages and times required under the Contract.
19. Ensure that a site diary and site records are properly kept and agreed where appropriate with the Contractor, and arrange for regular progress photographs to be taken.
20. Ensure that site meetings are held at least once a month and that minutes are kept and agreed.
21. Issue certificates for payments after interim measurements promptly (see the *ICE Conditions of Contract*, clause 60).
22. Visit the site regularly – at least once a month. Inspect works in progress and review compliance with the Contract programme.
23. Ensure that nominated subcontractors are properly appointed by the main contractor and that appropriate Subcontract Conditions of Contract are applied (see the *ICE Conditions of Contract*, clause 59).
24. Agree measurements of quantities for completed work as the work proceeds, and agree with the Contractor that they are to be carried unaltered to the final account (see the *ICE Conditions of Contract*, clause 48).
25. Ensure that claims are properly detailed and that any sums due thereon are settled as soon as possible (see the *ICE Conditions of Contract*, clause 52).
26. Ensure that Certificates of Maintenance and Completion are issued to the Contractor on time (see the *ICE Conditions of Contract*, clauses 48 and 61).
27. Ensure that the Employer is aware of his new insurance liability when the Maintenance Certificate is issued (see the *ICE Conditions of Contract*, clause 21).
28. When preparing decisions under clause 66 of the *ICE Conditions of Contract*, review all the evidence available and, if possible, arrange for the Engineer's Representative to put the Employer's case, and the Contractor his, to enable a clear

judgement to be made of the issues.

29. In the event of arbitration, try to keep the dispute within the scope of the Engineer's decision given under clause 66 of the *ICE Conditions of Contract*, and present evidence fairly and concisely.[22,23]

General

The guiding principle for the Engineer and his staff is that the Contract is a joint enterprise for the benefit of both parties: the Employer is entitled to a project well executed and the Contractor to fair dealing and a fair profit. It should always be remembered that the Contractor could only price and resource for the works which were defined at the time he tendered.

Appendix 2

Some standard forms of contract

Civil engineering

Association of Consulting Engineers *et al. Conditions of contract for overseas works mainly of civil engineering construction with forms of tender and agreement*. ACE, ICE and Export Group for the Constructional Industries, London, 1956.

Federation of Civil Engineering Contractors. *Form of subcontract designed for use in conjunction with the ICE Conditions of Contract*. FCEC, London, 1973, revised 1984.

Fédération Internationale des Ingénieurs-Conseils *et al. Conditions of contract (international) for works of civil engineering construction with forms of tender and agreement*, 3rd edn. 1977. Issued in London by Export Group for the Constructional Industries.

Institution of Civil Engineers *et al. Conditions of contract and forms of tender, agreement and bond for use in connection with works of civil engineering construction*, 5th edn. ICE *et al.*, London, 1986 revision.

Institution of Civil Engineers *et al. ICE conditions of contract for ground investigation*. Thomas Telford, London, 1983.

Civil engineering and building

Department of the Environment. *General conditions of government contracts for building and civil engineering minor works*. HMSO, London, 1980, Form GC/Works/2.

Department of the Environment. *General conditions of government contracts for building and civil engineering works*, 2nd edn. HMSO, London, 1977, Form GC/Works/1.

Department of the Environment. *General conditions of government contracts for building, civil engineering, mechanical and electrical small works*. HMSO, London, 1985, Form C 1001.

Building

Association of Consultant Architects. *Form of building agreement*.

British Property Federation. *System for building design and construction*. BPF, London, 1983.

Faculty of Architects and Surveyors. *Small works contract*. FAS, Chippenham, 1981.

Joint Contracts Tribunal. *Agreement for minor building works*. Royal Institute of British Architects, London, 1980.

Joint Contracts Tribunal. *Intermediate form of building contract*. Royal Institute of British Architects, London, 1984.

Joint Contracts Tribunal. *Standard fixed fee form of prime cost contract*. Royal Institute of British Architects, London, 1976.

Joint Contracts Tribunal. *Standard form of contract with approximate quantities*, Local authority and private editions. Royal Institute of British Architects, London, 1980.

Joint Contracts Tribunal. *Standard form of contract with contractors design*. Royal Institute of British Architects, London, 1981.

Joint Contracts Tribunal. *Standard form of contract with quantities*, Local authority and private editions. Royal Institute of British Architects, London, 1980.

Joint Contracts Tribunal. *Standard form of contract without quantities*, Local authority and private editions. Royal Institute of British Architects, London, 1980.

Joint Contracts Tribunal. *Standard form of management contract*. Royal Institute of British Architects, London. To be published.

Electrical and mechanical

Fédération Internationale des Ingénieurs-Conseils. *Conditions of contract (international) for electrical and mechanical works including erection on site*.

Institution of Electrical Engineers and Institution of Mechanical Engineers. *Model form A: Home contracts with erection. Model form B1: Export contracts for the supply of plant and machinery. Model form B2: Export contracts delivery FOB, CIF or FOR with supervision of erection. Model form B3: Export contracts including delivery to and erection on site. Model form C: Electrical and mechanical goods other than electric cables – home with erection. Model form E: Cable contracts with installation – home or export*. IEE and IMechE, London. To be published.

Process plants

Institution of Chemical Engineers. *Lump sum contracts in the United Kingdom*. IChemE, London, 1981.

Institution of Chemical Engineers. *Model form of conditions of contract for process plants suitable for reimbursable contracts in the United Kingdom*, IChemE, London, 1976.

Landscaping

Joint Council for Landscape Industries. *Agreement for landscape works*. Landscape Institute, London, 1985.

Appendix 3

Equivalent terms used in some standard forms of contract*

ICE[2] and FIDIC[12]	IMechE and IEE model form A[8]	GC/Works/1[26]	JCT building[27]
Employer	Purchaser	Authority	Employer
Contractor	Contractor	Contractor	Contractor
Engineer	Engineer	Superintending Officer	Architect or Supervising Officer
Works (Temporary and Permanent)	Works	Works (Permanent)	Works
Tender Total	Contract Price	Contract Sum	Contract Sum
Contract Price	Contract Value	Final Sum	Contract Sum adjusted
Interim Certificate	Interim Certificate	Certificate	Interim Certificate
Final Certificate	Final Certificate	Final Certificate	Final Certificate
Completion Certificate	Taking-over Certificate	Certificate	Certificate of Practical Completion
Maintenance Period	Defects Period after Taking Over	Maintenance Period	Defects Liability Period
Maintenance Certificate	Final Certificate	Certificate	Certificate of Completion of Making Good Defects

* In some cases terms are not exactly equivalent.

Appendix 4

Methods of programming

A simple precedence network and the linked bar chart prepared from it are shown in Figs 3 and 4 respectively. Fig. 5 shows the bar chart with progress marked on it.

Precedence network

The precedence network in Fig. 3 shows each item of construction for a reinforced concrete service reservoir contract represented by a box containing details of the description, duration and number of the item. The sequence of work is shown by arrows or relationships. These include time delays in some instances and illustrate the overlapping of some items. Where the start of an item is controlled by the start of another, or where the finish of an item is dependent on the finish of another, this is indicated by the position of the start and finish of the arrows as shown in the key.

Linked bar chart programme

Figure 4 shows the linked bar chart prepared from the precedence network in Fig. 3. It is a network displayed in bar chart form with hatched bars for eight items of construction with intended dates for start and completion. The total quantities to be executed and the average progress required each week to complete on time are also given. Average progress, however, is not generally practicable, as at first the rate is low, then it increases to a maximum and then it diminishes for the final cleaning up. The anticipated progress is shown as cumulative quantity targets above the programme bars. From this chart can be determined when plant, labour and materials for particular operations are required.

Programme with progress

Figure 5 shows the chart in Fig. 4 as it appears at the end of week 22 with progress entered. The progress line has been marked in. As each week passes actual quantities executed are blacked in on the programme bars according to the scale of the programme figures. At the same time the actual dates and working period are plotted as a separate time line below the bar. Actual quantities executed each week are shown in cumulative figures above the time line. The progress should always be plotted from the beginning of the pro-

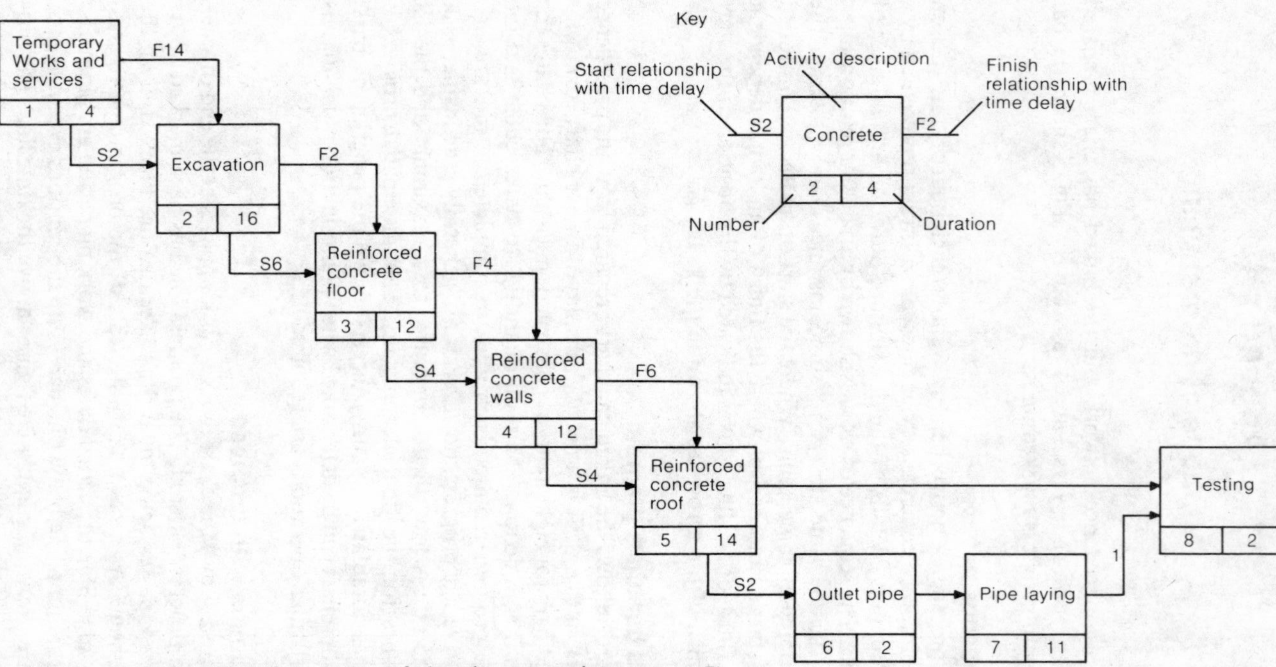

Fig. 3. Specimen precedence network (service reservoir: contract 7)

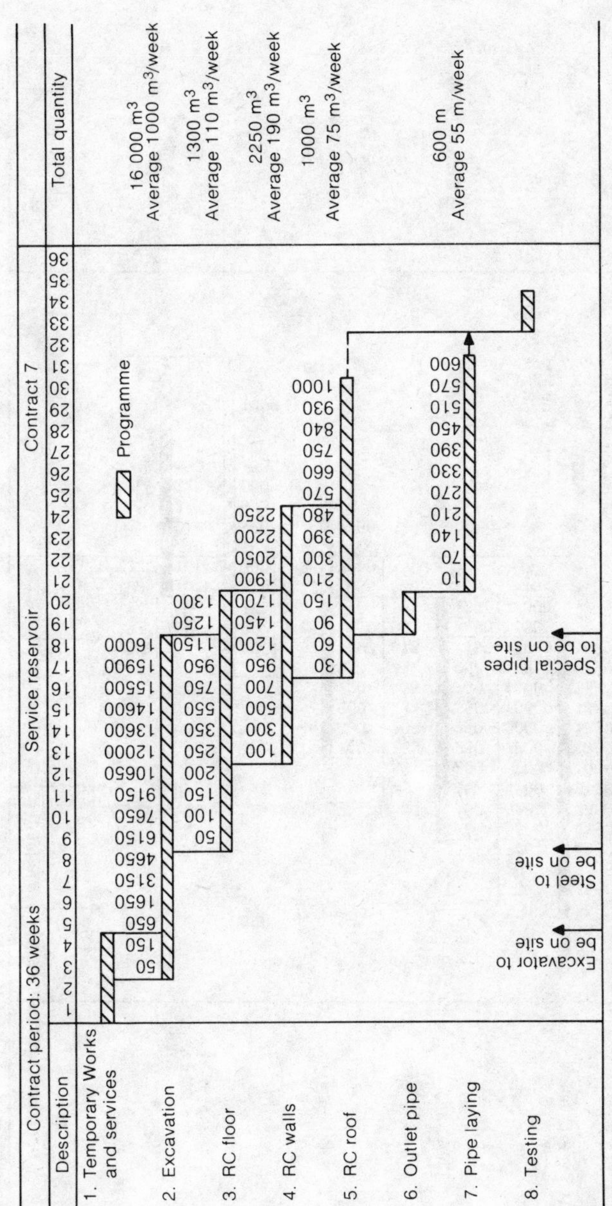

Fig. 4. Specimen linked bar chart programme

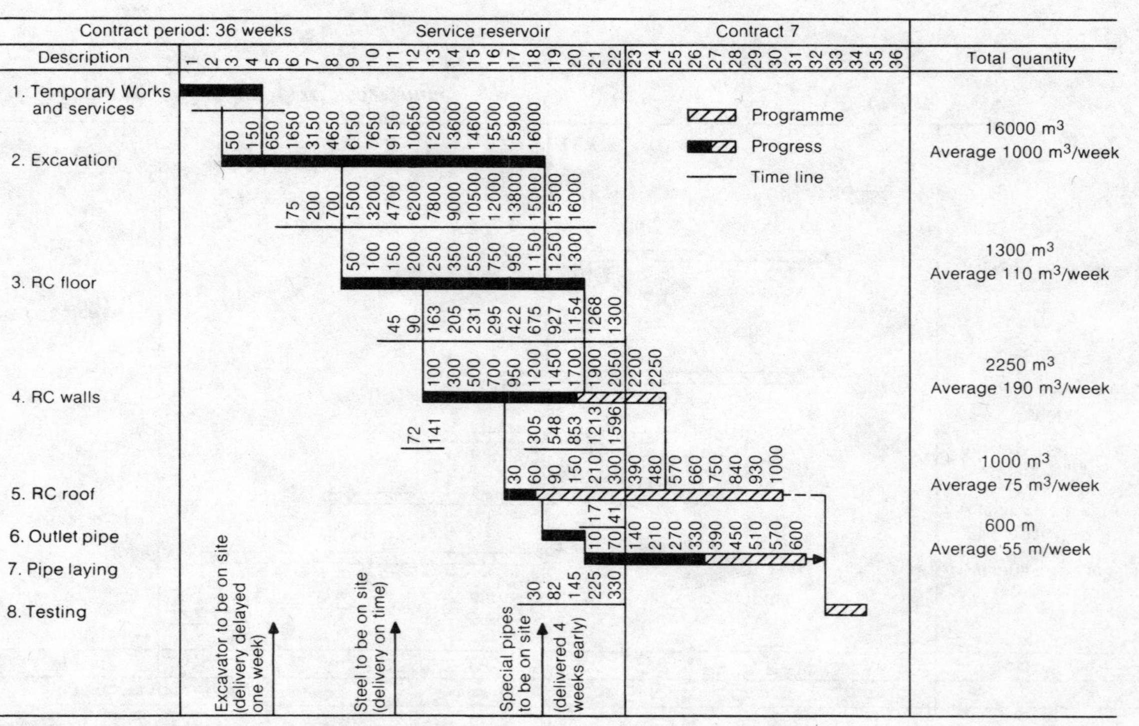

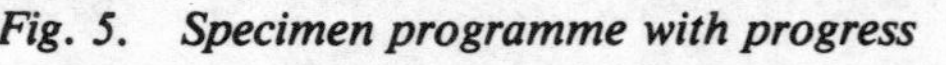
Fig. 5. Specimen programme with progress

gramme bar, regardless of when the actual work was begun. It is readily seen that operations 1, 2, 3 and 6 have been completed. Operations where progress is behind (i.e. where the blacked in programme band does not reach the progress line) are items 4 and 5. Progress for item 7 is ahead of programme. The position at the end of week 22 is as follows.

(*a*) The temporary works and services were started 1 week late and finished 2 weeks behind programme.

(*b*) The excavation was started 3 weeks late and finished 2 weeks behind programme.

(c) The reinforced concrete floor was started 1 week late and finished 2 weeks behind programme.

(*d*) The concreting cycle was altered to reduce the delay from the start of the floor to the start of the walls from 4 weeks to 1 week. As a result the concrete walls were started a week ahead of programme but progress was interrupted and is behind by 2½ weeks. Only 1596 m^3 have been done instead of 2050 m^3 as intended. To complete on programme 654 m^3 remain to be done in 2 weeks – which may be possible by working overtime.

(*e*) The concrete roof was started 4 weeks late and is 4¾ weeks behind programme. Only 41 m^3 have been done instead of 300 m^3 as intended. To complete 1000 m^3 according to programme, 959 m^3 remain to be done in 8 weeks. Shuttering sufficient for only 90 m^3 of concrete per week has been prepared, whereas at least 120 m^3 per week average will be required. Additional shuttering and overtime will be necessary if the roof is to be completed by the end of week 30, to give time for striking and cleaning up before testing.

(*f*) It proved possible to start the outlet pipe before the roof. The time line for the outlet pipe has been omitted but it is evident that since the pipelaying started 3 weeks early the outlet pipe must have been completed by at least 3 weeks early.

(*g*) The pipelaying is 4 weeks ahead of programme; 330 m are completed, leaving 270 m to be done.

Appendix 5

Some acts and regulations relevant to project design and construction in the UK

Arbitration Acts 1950, 1975 and 1979
Boiler Explosions Act 1882 and 1890 (Repeals and Modifications), Regulations 1974
Control of Pollution Act 1974
Employers' Liability (Compulsory Insurance) Act 1969
Employers' Liability (Defective Equipment) Act 1969
Employment Medical Advisory Service Act 1972
Employment Protection Act 1975
Employment Protection (Consolidation) Act 1978
Employment of Women, Young Persons and Children Act 1920
Explosives Act 1875 as amended by Explosives Act 1923 and (Exemptions) Regulations 1979
Factories Act 1961
- Factories Act (Docks, Building and Engineering Construction etc.) Modification Regulations 1938 (SR&O 1938, No. 610)
- Construction (General Provisions) Regulations 1961 (SI 1961, No. 1580)
- Construction (Lifting Operations) Regulations 1961 (SI 1961, No. 1581)
- Construction (Working Places) Regulations 1966 (SI 1966, No. 94)
- Construction (Health and Welfare) Regulations 1966 (SI 1966, No. 95 as amended by SI 1974, No. 209)
- Engineering Construction (Extension of Definition) Regulations 1960 (SI 1960, No. 421)
- Engineering Construction (Extension of Definition) Regulations 1968 (SI 1968, No. 1530)
- Work in Compressed Air Special Regulations 1958 (SI 1958, No. 61 as amended by SI 1960, No. 1307)
- Electricity Regulations 1908 (SR&O 1908, No. 1312) as amended by Electricity (Factories Act) Special Regulations 1944 (SR&O 1944, No. 739)
- Woodworking Machines Regulations 1974 (SI 1974, No. 903)
- Protection of Eyes Regulations 1974 (SI 1974, No. 1681)
- Ionising Radiations (Sealed Sources) Regulations 1969 (SI 1969, No. 808)

- Ionising Radiations (Unsealed Radioactive Substances) Regulations 1968 (SI 1968, No. 780)
- Asbestos Regulations 1969 (SI 1969, No. 690)
- Abrasive Wheels Regulations 1970 (SI 1970, No. 535)
- Highly Flammable Liquids and Liquefied Petroleum Gases Regulations 1972 (SI 1972, No. 917)

Fire Precautions Act 1971

Health and Safety at Work etc. Act 1974

- Industrial Tribunals (Improvement and Prohibition Notices Appeals) Regulations 1974, No. 1925
- Industrial Tribunals (Improvement and Prohibition Notices Appeals) (Scotland) Regulations 1974, No. 1926
- Factories Act 1961 etc. (Repeals and Modifications) Regulations 1974, No. 1941
- Petroleum (Regulation) Acts 1928 and 1936 (Repeals and Modifications) Regulations 1974, No. 1942
- Offices, Shops and Railway Premises Act 1963 (Repeals and Modifications) Regulations 1974, No. 1943
- Pipe-lines Act 1962 (Repeals and Modifications) Regulations 1974, No. 1986
- Mines and Quarries Acts 1954–71 (Repeals and Modifications) Regulations 1974, No. 2013
- Health and Safety Licensing Appeals (Hearings Procedure) Rules 1974, No. 2040
- Nuclear Installations Act 1965 etc. (Repeals and Modifications) Regulations 1974, No. 2056
- Health and Safety Licensing Appeals (Hearings Procedure) (Scotland) Rules 1974, No. 2068
- Explosives Acts 1875 and 1923 etc. (Repeals and Modifications (Amendment)) Regulations 1974, No. 2166
- Protection of Eyes (Amendment) Regulations 1975, No. 303
- Health and Safety Inquiries (Procedure) Regulations 1975, No. 335
- Offices, Shops and Railway Premises Act 1963 (Repeals) Regulations 1975, No. 1011
- Factories Act 1961 (Repeals) Regulations 1975, No. 1012
- Mines and Quarries Acts 1954 to 1971 (Repeals and Modifications) Regulations 1975, No. 1102
- Employers' Health and Safety Policy Statements (Exception) Regulations 1975, No. 1584
- Operations at Unfenced Machinery (Amendment) Regulations 1976, No. 955
- Health and Safety Inquiries (Procedure) (Amendment) Regulations 1976, No. 1246

Fire Certificates (Special Premises) Regulations 1976, No. 2003
Factories Act 1961 etc. (Repeals) Regulations 1976, No. 2004
Offices, Shops and Railway Premises Act 1963 etc. (Repeals) Regulations 1976, No. 2005
Fire Precautions Act 1971 (Modifications) Regulations 1976, No. 2007
Mines and Quarries (Metrication) Regulations 1976, No. 2063
Safety Representatives and Safety Committees Regulations 1977, No. 500
Health and Safety (Enforcing Authority) Regulations 1977, No. 746
Explosives (Registration of Premises) Variation of Fees Regulations 1977, No. 918
Health and Safety at Work etc. Act 1974 (Application outside Great Britain) Order 1977, No. 1232
Packaging and Labelling of Dangerous Substances Regulations 1978, No. 209
Explosives (Licensing of Stores) Variation of Fees Regulation 1978, No. 270
Petroleum (Regulation) Acts 1928 and 1936 (Variation of Fees) Regulations 1978, No. 635
Factories (Standards of Lighting) Revocation Regulations 1978, No. 1126
Compressed Acetylene (Importation) Regulations 1978, No. 1723
Mines and Quarries Act 1954 (Modification) Regulations 1978, No. 1951
Petroleum (Consolidation) Act 1928 (Enforcement) Regulations 1979, No. 427
Explosives Act 1875 (Exemptions) Regulations 1979, No. 1378
Notification of Accidents and Dangerous Occurrences Regulations 1980, No. 804
Health and Safety (Leasing Arrangements) Regulations 1980, No. 907
Agriculture (Tractor Cabs) (Amendment) Regulations 1980, No. 1036
Petroleum (Consolidation) Act 1928 (Conveyance by Road Regulations Exemptions) Regulations 1980, No. 1100
Mines and Quarries (Fees for Approvals) Regulations 1980, No. 1233
Control of Lead at Work Regulations 1980, No. 1248
Safety Signs Regulations 1980, No. 1471
Health and Safety (Enforcing Authority) (Amendment) Regulations 1980, No. 1744
Mines and Quarries (Fees for Approvals) (Amendment)

Regulations 1981, No. 270
Health and Safety (Fees for Medical Examinations) Regulations 1981, No. 334)
Diving Operations at Work Regulations 1981, No. 399
Pipe-lines Act 1962 (Metrication) Regulations 1981, No. 695
Packaging and Labelling of Dangerous Substances (Amendment) Regulations 1981, No. 792
Health and Safety (First-Aid) Regulations 1981, No. 917
Dangerous Substances (Conveyance by Road in Road Tankers and Tank Containers) Regulations 1981, No. 1059

Health and Safety Commission approved codes of practice
Control of lead at work (1980, revised 1981)
Work with asbestos insulation and asbestos coating (1981, revised 1983)
For the Health and Safety (First-Aid) Regulations 1981

Hours of Employment (Conventions) Act 1936

Mineral Workings (Offshore Installations) Act 1971
Offshore Installations (Diving Operations) Regulations 1974 (SI 1974, No. 1229)

Mines and Quarries Acts 1954–71

National Insurance (Industrial Injuries) Act 1965

Nuclear Installations Act 1965

Offices, Shops and Railway Premises Act 1963 as subsequently modified by Exemption Orders and Regulations

Pipe-lines Act 1962

Race Relations Act 1976

Reservoirs Act 1975

Sex Discrimination Act 1976

Trade Union and Labour Relations Act 1974

Appendix 6
Arbitration

Negotiation and settlement is usually to be preferred to arbitration, which should be regarded as a last resort. In those cases where arbitration seems inevitable, much time and expense will be saved if the necessary preliminary steps are properly understood and implemented.

Arbitration under the *ICE Conditions of Contract*[2] is covered in detail in *The Institution of Civil Engineers' arbitration practice*.[28] The preliminary steps are now outlined briefly.

Two versions of the ICE arbitration clause 66 are currently in use: the original version included in the fifth edition of the *ICE Conditions of Contract* published in 1973, and a revised version included in the fifth edition revision published in 1986.[2] In addition, some arbitrations under the fourth edition version may still arise. However, the principles behind all three versions are the same.

The dispute or difference

In almost every civil engineering project problems will arise at all stages of construction which the Engineer will deal with by means of instructions, variation orders or other decisions. Many of these problems will result in claims which the Engineer will accept or reject as the Works proceed. The Engineer's actions (or inactions) will not always be to the liking of the Contractor or of the Employer, but no 'dispute or difference' within the meaning of clause 66 will arise unless and until one party or the other decides to take a particular matter further. Thus the Contractor may not like the Engineer's decision on his claim or the Employer may be aggrieved at the granting of an extension of time, but until the aggrieved party decides to reject the Engineer's ruling there is no dispute or difference.

Reference to the Engineer

Once a dispute or difference has come into being in this way, it cannot be referred to arbitration under clause 66 until it has first been referred back to the Engineer for a formal decision under that clause. This is so whether or not he has already given a decision on the subject at an earlier stage.

However, provided that it is made clear that the reference is being made under clause 66, no special form of submission is necessary.

Thus, where a fully detailed case has been submitted at an earlier stage, a simple letter asking for a clause 66 decision thereon may suffice. Nevertheless, it is open to the applicant party to recast his claim – perhaps to incorporate further information or argument – or to base it on some other provision of the Contract.

From the Engineer's point of view, when a matter is referred to him under clause 66 he is bound to consider the matter afresh. In so doing, he may feel it right to reverse or vary any earlier decision that he may have given. For example, he may have rejected a claim earlier on the grounds that he was not satisfied that the claim had been substantiated, but when it returns to him for a clause 66 decision he must reassess the claim on its merits and come to a fair and balanced decision, even where it may be alleged that the claim concerns matters in which he or his staff have been at fault. In any event, the Engineer should attempt to give a decision which will assist the parties and, if he feels able also to state his reasons, so much the better.

It should be understood that, although the Engineer has a general duty to act fairly and impartially and to hold the balance between the parties in his administration of the Contract, he is not an arbitrator and to think of his functions under clause 66 as quasi-arbitral can be misleading. Thus, unlike a true arbitrator, the Engineer is not bound to hear or receive submissions from both parties before reaching his decisions (although he may often be well advised to do so). He should nevertheless be careful to avoid the appearance of acting merely as a rubber stamp, particularly where either party employs auditors* or other outside experts to monitor progress, or where the Employer's internal organization is such that pressure could be brought to bear on the Engineer or his staff.

Although it is important that the parties know when a clause 66 decision has been given, it is not open to the Engineer to stipulate that a particular decision is given under the clause, as the effect would be to deprive a party of his rights under the Contract.

Reference to arbitration

Once the Engineer has given his decision under clause 66 – or has failed to do so within the time laid down in that clause – a party who is aggrieved at the decision (which need not be the one who made the clause 66 submission) may refer the dispute or decision to arbitration. This must be done within a further prescribed period, but need not follow any particular form, provided that due and

* The correct relationship between the Engineer and an auditor is set out in reference 1.

timely notice is given to the other party.† On the giving of such notice arbitration is deemed to commence. It should be emphasized that, if notice is not given within the prescribed time, the Engineer's decision becomes final and binding on both parties and neither can thereafter challenge the decision, either by arbitration or through the courts.

Time limits

Once a dispute or difference has arisen, it may be referred to the Engineer for a clause 66 decision at any convenient time, whether during the progress of the Works, after substantial completion or even after the final account has been submitted, and the Engineer has a fixed time within which he must give his decision. Under the fourth edition of the *ICE Conditions of Contract* and under the 1973 fifth edition version of clause 66 that period was three calendar months from the date of the reference. However, under the 1986 revision a distinction is drawn between requests submitted before substantial completion of the whole of the Works – when the period is one calendar month – and those submitted after such completion – when the period continues to be three calendar months. The object of the shorter period is to ensure that disputes during the currency of the Works are dealt with expeditiously; the longer period is retained to cover requests made after the Engineer's staff may have been dispersed to other projects. In any event, the Engineer should always give as quick a decision as may be consonant with proper consideration, and should not treat the prescribed period as a justification for delay.

Once the Engineer has given his decision – or, if he fails to give one, when the prescribed period has expired – the parties have a further three calendar months within which to give notice of referral to arbitration under all three versions of clause 66. Failure to give notice in time will normally prevent the issue from being contested further.

Interim arbitration

Under the version of clause 66 appearing in the fourth edition of the *ICE Conditions of Contract*, no steps in the arbitration beyond the giving of notice of referral thereto and the appointment of a suitable arbitrator could take place until after the completion or alleged completion of the Works unless both the Contractor and the Employer gave their written consent. The only exceptions were

† Forms for this purpose are obtainable from the Institution of Civil Engineers' Arbitration Officer.

disputes about the withholding by the Engineer of any certificate, the withholding of any portion of the retention money payable under clause 60, or the exercise of the Engineer's power to give a certificate under clause 63(1) as to forfeiture, when in each case the arbitration could proceed at any time. Similar provisions appeared under the 1973 version of clause 66 in the fifth edition, except that any matter arising under clause 12 was added to the list of exceptions and an arbitration under any of these exceptions was termed an 'Interim Arbitration' in the marginal note, although not in the text.

However, under the 1986 revision these distinctions are not made and any reference to arbitration may now proceed before substantial completion of the Works unless both the Contractor and the Employer agree in writing that it should not. The former rule is thus now reversed, and all arbitrations can be interim arbitrations, and thereby benefit from the accelerated procedures provided in the Institution of Civil Engineers' *Arbitration procedure (1983)*[22,23] – which itself is now mandatory whereas before it was only optional.

Bibliography

Contracts

Abrahamson M.W. *Engineering law and the ICE contracts*, 4th edn. Applied Science, London, 1975.

Atkinson A.V. *Civil engineering contract administration*, 1st edn. Hutchinson, London, 1985.

Duncan Wallace I.N. *The ICE Conditions of Contract 5th edn. A commentary*. Sweet & Maxwell, London.

Horgan M.O. *Competitive tendering for engineering contracts*. E&F Spon, London, 1984.

Horgan M.O. and Roulston F.R. *The elements of engineering contracts*. W.S. Atkins, Epsom, 1977.

Hudson's building and engineering contracts, 10th edn. Sweet & Maxwell, London, 1970.

Institution of Civil Engineers. *Guidance note on the Engineer as expert witness*. ICE, London.

Keating D. *Building contracts*, 4th edn. Sweet & Maxwell, London, 1978, with 1982–84 supplements.

Marks R.J. *et al. Aspects of civil engineering contract procedure*, 2nd edn. Pergamon, Oxford, 1978.

Major W.T. and Ranson A. *Building and engineering claims*. Oyez, London, 1980.

Uff J.F. *Construction law*, 4th edn. Sweet & Maxwell, London, 1985.

Management

Abrahamson M.W. Risk management. *Int. Constr. Law Rev.*, 1984, **1**, Part 3.

Burman P.J. *Precedence networks for project planning and control*. McGraw-Hill, Maidenhead, 1972.

Clarke R.H. *Site supervision*. Thomas Telford, London, 1984.

Cooper B. *Writing technical reports*. Penguin, Harmondsworth, 1978.

Goodlad J.B. *Accounting for construction management*. Heinemann, London, 1974.

Harris F. and McCaffer R. *Building modern construction management*. Crosby Lockwood Staples – Granada, St Albans, 1978.

Institution of Civil Engineers. *Supervision of construction*. ICE, London, 1984.

Institution of Civil Engineers. *Management of large capital projects*. ICE, London, 1978.

Kennaway A. Errors and failures in building – why they happen and what can be done about them. *Int. Constr. Law Rev.*, 1984, **2**, Part 1.

Mitchell J. *How to write reports*. Fontana, London, 1974.

Macmillan of Aberfeldy. *The giving of evidence before a parliamentary committee, in the High Court, and before an arbitrator*. ICE, London, 1975.

National Economic Development Office. *Faster building for industry*. HMSO, London, 1983.

National Economic Development Office. *The public client and the construction industries*. HMSO, London, 1975.

Ninos G.E. and Wearne S.H. *Responsibilities for project control during construction*. School of Technological Management, University of Bradford, 1984, report TMR 17.

Scott W. *Communication for professional engineers*. Thomas Telford, London, 1984.

Thompson P.A. *Organisation and economics of construction*. McGraw-Hill, Maidenhead, 1981.

Measurement

Barnes M. *The CESMM2 handbook*. ICE, London, 1986.

Department of Transport. *Method of measurement for road and bridge works*, 2nd edn. HMSO, London, 1977.

Department of Transport. *Notes for guidance and library of standard item descriptions for the preparation of Bills of Quantities for road and bridge works*. HMSO, London, 1978.

McCaffer R. and Boldwin A.N. *Estimating and tendering for civil engineering works*. Granada, London, 1984.

Seely I.H. *Civil engineering quantities*, 3rd edn. Macmillan, London, 1977.

Legislation

Allsop P. and Goodman M.J. *Encyclopedia of health and safety at work, law and practice*. Sweet & Maxwell, London, 1966 with updating.

Construction Industry Research and Information Association. *Building design legislation: a guide to the acts of Parliament and government orders and regulations which affect the design of buildings in England and Wales*. CIRIA, London, 1982, SP 23.

Construction Industry Research and Information Association. *Scottish building legislation*. CIRIA, London, 1985, SP 34.

Fife I. and Machin E.A. *Health and safety at work*. Butterworths, London, 1980.

Fife I. and Machin E.A. *Redgrave's health and safety in factories*, 2nd edn. Butterworths, London, 1982.

Arbitration

Hawker G. *et al. The Institution of Civil Engineers' arbitration practice*. Thomas Telford, London, 1986.

Mildred R.H. *The expert witness*. George Godwin, London, 1982.

Mustill M.J. and Boyd S.L. *The law and practice of commercial arbitration in England*. Butterworths, London, 1982.

Walton A. and Vitoria M. *Russell on the law of arbitration*, 20th edn. Stevens, London, 1982.

Insurance

Eaglestone F.N. and Smyth C. *Insurance under the ICE contract*. George Godwin, London, 1985.

Madge P. *A guide to the indemnity and insurance aspects of building contracts*. Royal Institute of British Architects, London, 1985.

References

1. Institution of Civil Engineers and Chartered Institute of Public Finance and Accountancy. *Joint statement – engineers and auditors*. ICE, London, 1983.
2. Institution of Civil Engineers *et al*. *Conditions of contract and forms of tender, agreement and bond for use in connection with works of civil engineering construction*, 5th edn. ICE *et al*. London, 1986 revision.
3. Association of Consulting Engineers. *Conditions of engagement*. ACE, London, 1963, 1970 and 1981.
4. Institution of Civil Engineers. *An introduction to engineering economics*. ICE, London, 1969.
5. Perry J.G. *et al*. *Target and cost reimbursible construction contracts*. CIRIA, London, 1980, Report R85, Part C.
6. Walker M.J. (ed.). *Management contracts*, CIRIA, London, 1983, Report R100.
7. Working Party on Direct Labour Organisations. *Final report*. Department of the Environment, London, 1978.
8. Institution of Electrical Engineers and Institution of Mechanical Engineers. *Model form A: Home contracts with erection. Model form B1: Export contracts for the supply of plant and machinery. Model form B2: Export contracts delivery FOB, CIF or FOR with supervision of erection. Model form B3: Export contracts including delivery to and erection on site. Model form C: Electrical and mechanical goods other than electric cables – home with erection. Model form E: Cable contracts with installation – home or export*. IEE and IMechE, London. To be published.
9. Institution of Civil Engineers. *Conditions of contract for ground investigation*. Thomas Telford, London, 1983.
10. Institution of Civil Engineers. *Civil engineering standard method of measurement*, 2nd edn. Thomas Telford, London, 1985.
11. Institution of Civil Engineers. *Guidance on the preparation, submission and consideration of tenders for civil engineering contracts, recommended for use in the United Kingdom*. ICE, London, 1983.
12. Fédération Internationale des Ingénieurs-Conseils *et al*. *Condi-*

tions of contract (international) for works of civil engineering construction with forms of tender and agreement, 3rd edn. 1977. Issued in London by Export Group for the Constructional Industries.

13. Civil Engineering Construction Conciliation Board for Great Britain. *Working rule agreement*. Federation of Civil Engineering Contractors, London, 1986.
14. British Standards Institution. *Use of network techniques in project management*, Parts 1–4. BSI, London, 1984, BS 6046.
15. British Standards Institution. *Glossary of terms used in project network techniques*. BSI, London, 1972, BS 4335.
16. Institution of Civil Engineers. Engineering for safety. *Proc. Instn Civ. Engrs*, Part 1, 1986, **80**, Feb., 13–119.
17. Institution of Civil Engineers. *Hazards in construction*. ICE, London, 1972.
18. Institution of Civil Engineers. *Hazards in tunnelling and on falsework*. ICE, London, 1975.
19. Institution of Civil Engineers. *Safety in wells and boreholes*. ICE, London, 1972.
20. Federation of Civil Engineering Contractors. *Supervisors' safety booklet*. FCEC, London, 1980.
21. National Joint Health and Safety Committee for the Water Services. *Safe work in sewers and at sewage works*. NJHSCWS, London, 1979, Health and Safety Guideline 2.
22. Institution of Civil Engineers. *Arbitration procedure (England and Wales) (1983)*. ICE, London, 1983.
23. Institution of Civil Engineers. *Arbitration procedure (Scotland) (1983)*. ICE, London, 1983.
24. Institution of Civil Engineers' Conditions of Contract Standing Joint Committee. *Functions of the Engineer under the ICE Conditions of Contract*. ICE, London, 1977, Guidance note 2A.
25. Institution of Civil Engineers' Conditions of Contract Standing Joint Committee. *Guidance on the preparation, submission and consideration of tenders for civil engineering contracts in the United Kingdom*. ICE, London, 1983.
26. Department of the Environment. *General conditions of government contracts for building and civil engineering works*, 2nd edn. HMSO, London, 1977, Form GC/Works/1.
27. Joint Contracts Tribunal. *Standard form of building contract*. Royal Institute of British Architects, London, 1980.
28. Hawker G. *et al*. *The Institution of Civil Engineers' arbitration practice*. Thomas Telford, London, 1986.

£6-00

Civil engineering procedure